RIEMANN-FINSLER
GEOMETRY

NANKAI TRACTS IN MATHEMATICS

Series Editors: Yiming Long and Weiping Zhang
Nankai Institute of Mathematics

Published

Nankai Tracts in Mathematics – Vol. 6

RIEMANN-FINSLER GEOMETRY

Shiing-Shen Chern

Nankai Institute of Mathematics
P. R. China

Zhongmin Shen

Indiana University Purdue University Indianapolis
USA

 **World Scientific**

NEW JERSEY · LONDON · SINGAPORE · BEIJING · SHANGHAI · HONG KONG · TAIPEI · CHENNAI

Published by

World Scientific Publishing Co. Pte. Ltd.
5 Toh Tuck Link, Singapore 596224
USA office: 27 Warren Street, Suite 401-402, Hackensack, NJ 07601
UK office: 57 Shelton Street, Covent Garden, London WC2H 9HE

Library of Congress Cataloging-in-Publication Data
Chern, Shiing-Shen 1911–2004
  Riemann-Finsler geometry / S.S. Chern, Zhongmin Shen.
   p. cm. -- (Nankai tracts in mathematics ; v. 6)
  Includes bibliographical references and index.
  ISBN 981-238-357-3 (alk. paper) -- ISBN 981-238-358-1 (pbk. : alk. paper)
   1. Finsler spaces. 2. Geometry, Riemannian. I. Shen, Zhongmin, 1963–   II. Title. III.
Series.

  QA689.C48 2005
  516.3'75--dc22

                                               2005040818

British Library Cataloguing-in-Publication Data
A catalogue record for this book is available from the British Library.

Printed in Singapore by World Scientific Printers (S) Pte Ltd

Preface

Two years ago David Bao, Zhongmin Shen and I published a book on Riemann-Finsler geometry through Springer Verlag. Riemann-Finsler geometry is not a generalization of Riemannian geometry. Riemann knew and began with the general case. He saw the main features of Riemannian geometry and remarked that the general case does not involve new ideas.

In this assessment he was only partially correct. It certainly cannot include global problems. In local problems these are also subtleties, which need manipulation.

The aim of this book is to provide an elementary account of Finsler geometry to show that Finsler geometry is essentially not more difficult. Such an account is desirable, as Finsler metrics have come up in many applications.

I am ashamed to say that Shen actually wrote the whole book, although the idea of the book originated from me. The manuscript was written at least five times and I am impressed by its clarity and simplicity. I went through it in a seminar but do not wish to evade any responsibility if mistakes are found.

S.S. Chern

January 2003

At the end of 2001, I visited S.S. Chern at Nankai Institute of Mathematics in Tianjin, P.R. China. During my visit, Chern told me that he wished to have a concise book written for graduate students and young geometers who are interested in Riemann-Finsler geometry. The primary goal is to introduce some basic concepts, examples, theorems and to bring the readers to the most current research areas in Finsler geometry. We had a thorough discussion on topics and I started to collect some materials for the manuscript. Soon I realized that it is very difficult to write such a book. There is no simple proof for some important examples and theorems. Often times, the computation has to be carried out using a computer program such as MAPLE.

Since the winter of 2001, I have been meeting Chern twice a year at Nankai Institute of Mathematics to work on our book project. The first draft of the manuscript was completed in the summer of 2002. However, I continued to make changes based on my discussion with Chern and comments from our colleagues. Chern wanted to check all the details by himself. Thus the submission of the manuscript was postponed for several times.

On December 3, I was stricken with the saddest news from my colleagues that S.S. Chern was no longer with us. Personally, I lost a great advisor. Without Chern's support and encouragement throughout the last decade, I would not have done any work in Finsler geometry.

I would like to take this opportunity to thank X. Chen, L. Kozma, X. Mo, C. Robles, H. Shimada, and G. O. Yildirim for their valuable comments.

Zhongmin Shen
December 2004

Contents

RIEMANN-FINSLER GEOMETRY

Chapter 1

Finsler Metrics

To measure the length of a smooth curve C parametrized by a map $c = c(t)$, $a \leq t \leq b$, in a manifold M, it suffices to define a nonnegative scalar function $F(x, \cdot)$ on every tangent space $T_x M$. Then the length of C is defined by

$$\mathcal{L}_F(C) = \int_a^b F\Big(c(t), \dot{c}(t)\Big) dt.$$

It is required that $\mathcal{L}_F(C)$ be independent of parametrization. F must be positively homogeneous with degree one,

$$F(x, \lambda y) = \lambda F(x, y), \qquad \lambda > 0.$$

The length structure induces a nonnegative function $d : M \times M \to [0, \infty)$ by

$$d(p, q) := \inf_C L(C),$$

where the infimum is taken over all smooth curves C from p to q. In general, d is irreversible, i.e., $d(p, q) \neq d(q, p)$ for some pairs of points $\{p, q\}$. It is required that F uniquely determine d_F. We impose a convexity condition on F, that is,

$$F(x, y_1 + y_2) \leq F(x, y_1) + F(x, y_2), \qquad y_1, y_2 \in T_x M. \tag{1.1}$$

Thus $F_x(\cdot) := F(x, \cdot)$ is a "norm" on $T_x M$ without the reversibility. In order to apply calculus to study the geometric properties of F, we assume that F is differentiable on $TM \setminus \{0\}$. Further, we replace the above convexity condition with a stronger convexity condition, that is, the Hessian,

1

$[F^2]_{y^i y^j}(x, y)$, is positive definite for any $y \in T_x M \setminus \{0\}$. This strong convexity implies the inequality (1.1). Therefore F_x is still a "norm" on $T_x M$. Such a "norm" is called a *Minkowski norm*. A scalar function F with the above properties is called a *Finsler metric*. When the norm F_x at every point $x \in M$ is *Euclidean*, the Finsler metric F is called an *Riemannian* metric.

Riemann-Finsler geometry is to study the geometric properties of Finsler metrics on a manifold. Intuitively, the Minkowski norm F_x or its unit ball $\mathcal{U}_x := \{F_x(y) < 1\}$ at each point x is an infinitesimal color pattern and it varies over the manifold. Thus a Finsler manifold is a "colorful" curved space. We study not only the curvature, but also the colors on the space.

1.1 Minkowski Norms

A Finsler metric on a manifold consists of the so-called Minkowski norms on every tangent space. Thus we first study the geometry of Minkowski norms on a vector space.

Let V be a finite dimensional vector space. A function $F = F(y)$ on V is called a *Minkowski norm* if it has the following properties:

(a) $F(y) \geq 0$ for any $y \in V$, and $F(y) = 0$ if and only if $y = 0$;
(b) $F(\lambda y) = \lambda F(y)$ for any $y \in V$ and $\lambda > 0$;
(c) F is C^∞ on $V \setminus \{0\}$ such that for any $y \in V$, the following bilinear symmetric functional $\mathbf{g}_y$ on V is an inner product,

$$\mathbf{g}_y(u, v) := \frac{1}{2} \frac{\partial^2}{\partial s \partial t} \Big[F^2(y + su + tv) \Big]_{s=t=0}.$$

The inner product $\mathbf{g}_y$ is called the *fundamental form* in the direction y. The pair (V, F) is called a *Minkowski space*. A Minkowski norm F is said to be *reversible* if $F(-y) = F(y)$ for $y \in V$.

Given a Minkowski space (V, F), let

$$S_F := \Big\{ y \in V \mid F(y) = 1 \Big\}.$$

S_F is a closed hypersurface around the origin, which is diffeomorphic to the standard sphere $S^{n-1} \subset R^n$. S_F is called the *indicatrix* of F.

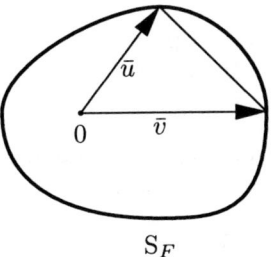

$$S_F$$

Figure 1.1

Let $u, v \in V \setminus \{0\}$ and $\bar{u} := u/F(u)$, $\bar{v} := v/F(v)$. Assume that $\bar{u} \neq \pm\bar{v}$. Consider

$$\varphi(t) := F^2(t\bar{u} + (1 - t)\bar{v}).$$

We have that $\varphi(0) = \varphi(1) = 1$ and

$$\varphi''(t) = 2\mathbf{g}_y(\bar{u} - \bar{v}, \bar{u} - \bar{v}) > 0, \qquad 0 < t < 1,$$

where $y := t\bar{u} + (1 - t)\bar{v}$. Thus $\varphi = \varphi(t)$ is strictly convex on $[0, 1]$. By a well-known result in calculus, $\varphi(t) < 1$, $0 < t < 1$, that is,

$$F(t\bar{u} + (1 - t)\bar{v}) < 1, \qquad 0 < t < 1.$$

Clearly, when $\bar{u} = -\bar{v}$, the above inequality still holds. Plugging $t = F(u)/(F(u) + F(v))$ into the above inequality yields

$$F(u + v) < F(u) + F(v).$$

In the case when $\bar{u} = \bar{v}$, the following equality holds

$$F(u + v) = F(u) + F(v).$$

Let (V, F) be a Minkowski space. Fix a basis $\{\mathbf{b}_i\}$ for V, and view $F(y) = F(y^i \mathbf{b}_i)$ as a function of $(y^i) \in R^n$. Then for $y \neq 0$,

$$g_{ij}(y) := \mathbf{g}_y(\mathbf{b}_i, \mathbf{b}_j) = \frac{1}{2}[F^2]_{y^i y^j}(y).$$

Here $[F^2]_{y^i y^j}(y)$ denote the partial derivative of F^2 with respect to y^i and y^j. We have

$$\mathbf{g}_y(u, v) = g_{ij}(y)u^i v^j, \qquad u = u^i \mathbf{b}_i, \; v = v^j \mathbf{b}_j$$

and

$$F(y) = \sqrt{g_{ij}(y)y^i y^j}, \qquad y = y^i \mathbf{b}_i.$$

Let us take a look at some special Minkowski norms. Let $\langle \, , \, \rangle$ be an inner product on a vector space V with a basis $\{\mathbf{b}_i\}$, and let

$$\alpha := \sqrt{\langle y, y \rangle} = \sqrt{a_{ij}y^i y^j}, \qquad y = y^i \mathbf{b}_i,$$

where $a_{ij} = \langle \mathbf{b}_i, \mathbf{b}_j \rangle$. Clearly, α is a Minkowski norm with $\mathbf{g}_y(u, v) = \langle u, v \rangle$ independent of $y \in V \setminus \{0\}$. α is called an *Euclidean norm*, and the pair (V, α) is called an *Euclidean space*. In each dimension, all Euclidean spaces are linearly isometric to each other. The standard Euclidean norm $|\cdot|$ on $\mathbf{R}^n$ is defined by

$$|y| := \sqrt{\sum_{i=1}^{n}(y^i)^2}, \qquad y = (y^i) \in \mathbf{R}^n.$$

There are many interesting non-Euclidean norms on a vector space. Let $\alpha = \sqrt{a_{ij}y^i y^j}$ be an Euclidean norm on a vector space V and $\beta = b_i y^i \in V^*$ be a linear functional on V. Let

$$F := \alpha(y) + \beta(y). \tag{1.2}$$

By a direct computation, one obtains

$$g_{ij} = \frac{F}{\alpha}\left\{a_{ij} - \frac{y_i}{\alpha}\frac{y_j}{\alpha} + \frac{\alpha}{F}\left(b_i + \frac{y_i}{\alpha}\right)\left(b_j + \frac{y_j}{\alpha}\right)\right\}, \tag{1.3}$$

where $y_i := a_{ij}y^j$. From (1.3), one can see that (g_{ij}) is positive definite if and only if the length of β is less than 1, i.e.,

$$\|\beta\|_\alpha := \sqrt{a^{ij}b_i b_j} < 1,$$

where $(a^{ij}) := (a_{ij})^{-1}$. A Minkowski norm in the form (1.2) is called the *Randers norm* ([79]).

For further computation, we need the following lemma from linear algebra.

Lemma 1.1.1 *Let $G = (g_{ij})$ and $H = (h_{ij})$ be symmetric $n \times n$ matrices and $C = (c_i)$ be an n-vector. Assume that H is invertible with $H^{-1} = (h^{ij})$,* *and*

$$g_{ij} = h_{ij} + \delta c_i c_j.$$

Then

$$\det(g_{ij}) = (1 + \delta c^2) \det(h_{ij}),$$

where $c := \sqrt{h^{ij} c_i c_j}$. *If* $1 + \delta c^2 \neq 0$, *then* G *is invertible. The inverse matrix* $G^{-1} = (g^{ij})$ *is given by*

$$g^{ij} = h^{ij} - \frac{\delta c^i c^j}{1 + \delta c^2},$$

where $c^i := h^{ij} c_j$.

Applying Lemma 1.1.1 to (1.3), we obtain

$$\det(g_{ij}) = \left(\frac{\alpha + \beta}{\alpha}\right)^{n+1} \det(a_{ij}). \tag{1.4}$$

By (1.4), we can also show that (g_{ij}) is positive definite if and only if $\alpha(y) + \beta(y) > 0$ for any $y \in V \setminus \{0\}$, if and only if $\|\beta\|_\alpha < 1$.

Using an Euclidean norm $\alpha = \sqrt{a_{ij} y^i y^j}$ and a 1-form $\beta = b_i y^i$ on a vector space V, one can define more general Minkowski norms—the (α, β)-*norms*:

$$F = \alpha \phi\left(\frac{\beta}{\alpha}\right). \tag{1.5}$$

Here the function $\phi = \phi(s)$ is a C^∞ positive function on some symmetric open interval $I = (-b_o, b_o)$. It is easy to see that $F = \alpha\phi(\beta/\alpha)$ is positively homogeneous of degree one. Let us find the condition for the positivity of $g_{ij} := \frac{1}{2}[F^2]_{y^i y^j}$. Assume that $b := \|\beta\|_\alpha < b_o$. Using a Maple program, one can easily compute the matrix (g_{ij}):

$$g_{ij} = \rho a_{ij} + \rho_0 b_i b_j + \rho_1 (b_i \alpha_j + b_j \alpha_i) - s\rho_1 \alpha_i \alpha_j,$$

where $\alpha_i = \alpha_{y^i}$ and

$$\rho = \phi^2 - s\phi\phi', \qquad \rho_0 = \phi\phi'' + \phi'\phi',$$

$$\rho_1 = -s(\phi\phi'' + \phi'\phi') + \phi\phi',$$

where the functions are evaluated on $s := \beta/\alpha$ with $|s| \leq b < r$. By Lemma 1.1.1, we find a formula for $\det(g_{ij})$.

$$\det(g_{ij}) = \phi^{n+1}(\phi - s\phi')^{n-2}\left[(\phi - s\phi') + (b^2 - s^2)\phi''\right]\det(a_{ij}).$$

Lemma 1.1.2 $F = \alpha\phi(\beta/\alpha)$ *is a Minkowski norm for any Riemannian metric* α *and 1-form* β *with* $\|\beta\|_\alpha < b_o$ *if and only if* $\phi = \phi(s)$ *satisfies the following conditions:*

$$\phi(s) > 0, \qquad (\phi(s) - s\phi'(s)) + (b^2 - s^2)\phi''(s) > 0, \qquad (1.6)$$

where s *and* b *are arbitrary numbers with* $|s| \leq b < b_o$.

Proof. Assume that (1.6) is satisfied. Then by taking $b = s$ in (1.6), we see that the following inequality holds for any s with $|s| < b_o$,

$$\phi(s) - s\phi'(s) > 0. \qquad (1.7)$$

Consider the following family of functions,

$$\phi_t(s) := 1 - t + t\phi(s).$$

Let $F_t := \alpha\phi_t(\beta/\alpha)$ and $g_{ij}^t := \frac{1}{2}[F_t^2]_{y^i y^j}(y)$. Note that for any $0 \leq t \leq 1$ and any s, b with $|s| \leq b < b_o$,

$$\phi_t - s\phi_t' = 1 - t + t\left[\phi - s\phi'\right] > 0,$$

$$(\phi_t - s\phi_t') + (b^2 - s^2)\phi_t'' = 1 - t + t\left[(\phi - s\phi') + (b^2 - s^2)\phi''\right] > 0.$$

Thus $\det(g_{ij}^t) > 0$ for all $0 \leq t \leq 1$. Since (y_{ij}^o) is positive definite, we conclude that (g_{ij}^t) is positive definite for any $t \in [0, 1]$. Therefore, F_t is a Minkowski norm for all $t \in [0, 1]$.

Conversely, assume that $F = \alpha\phi(\beta/\alpha)$ is a Minkowski norm for any Riemannian metric α and 1-form β with $b := \|\beta\|_\alpha < b_o$. Then $\phi(s) > 0$ for any s with $|s| < b_o$. If $n = even$, then $\det(g_{ij}) > 0$ implies that (1.6) holds for any s with $|s| \leq b$. If $n = odd > 1$, then $\det(g_{ij}) > 0$ implies that the following inequality holds for any s with $|s| \leq b$,

$$\phi(s) - s\phi'(s) \neq 0.$$

Since $\phi(0) > 0$, the above inequality implies that the inequality (1.7) holds for any s with $|s| \leq b$. Since the number b can be arbitrary with $0 \leq b < b_o$, we conclude that (1.7) holds for any s with $|s| < b_o$. Finally, we can see that $\det(g_{ij}) > 0$ implies that (1.6) holds for any s and b with $|s| \leq b < b_o$.

Q.E.D.

Sabau–Shimada have studied certain (α, β)-norms and they have also computed the Hessian g_{ij} for these metrics [83].

Let us take a look at some special (α, β)-norms. Let $\alpha = \sqrt{a_{ij} y^i y^j}$ and $\beta = b_i y^i$ be a Euclidean norm and a 1-form on a vector space V, respectively. Let $y_i := a_{ij} y^j$ and $b := \|\beta\|_\alpha$. Then $|\beta(y)| \leq b\alpha(y)$ for any $y \in$ V. Let

$$\phi := 1 + \varepsilon s + k s^2,$$

where ε and k are constants. The (α, β)-norm defined by ϕ is given by

$$F = \alpha + \varepsilon \beta + k \frac{\beta^2}{\alpha}.$$

By the above formulas, one obtains

$$g_{ij} = \frac{(\alpha^2 - k\beta^2)F}{\alpha^3} a_{ij} + \frac{6kF + (\varepsilon^2 - 4k)\alpha}{\alpha} b_i b_j$$
$$+ \frac{\varepsilon\alpha^3 - 3\varepsilon k\alpha\beta^2 - 4k^2\beta^3}{\alpha^4} \left\{ (b_i y_j + b_j y_i) - \frac{\beta}{\alpha^2} y_i y_j \right\},$$

and

$$\det(g_{ij}) = \left(\frac{\alpha^2 - k\beta^2}{\alpha^3} \right)^n F^{n+1} \frac{[(1 + 2kb^2)\alpha^2 - 3k\beta^2]\alpha}{(\alpha^2 - k\beta^2)^2} \det(a_{ij}).$$

Observe that

$$\phi(s) - s\phi'(s) + (b^2 - s^2)\phi''(s) = 1 + 2kb^2 - 3ks^2.$$

By Lemma 1.1.2, F is a Minkowski norm for any α and β with $\|\beta\|_\alpha < b_o$ if and only if

$$1 + \varepsilon s + k s^2 > 0, \qquad 1 + 2kb^2 - 3ks^2 > 0,$$

for any numbers s and b with $|s| \leq b < b_o$. Where $\varepsilon = 2$ and $k = 1$, F can be expressed as

$$F = \frac{(\alpha + \beta)^2}{\alpha}.$$

Thus this function is a Minkowski norm if $\|\beta\|_\alpha < 1$.

Now we are going to construct Minkowski norms by shifting a Minkowski norm. Let (V, Φ) be a Minkowski space and let $v \in V$ with $\Phi(-v) < 1$. Then the shifted set, $S_\Phi + \{v\}$, contains the origin of V.

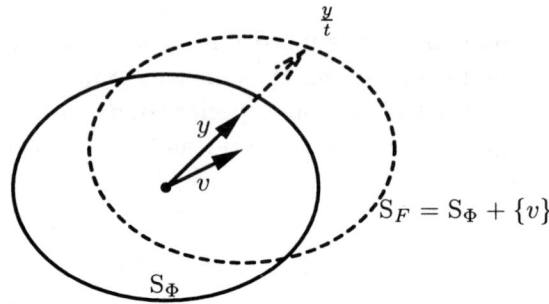

Figure 1.2

We can define a function $F : V \to [0, \infty)$ as follows: for any $y \in V \setminus \{0\}$, $F(y)$ is the unique positive number $t > 0$ such that

$$\frac{y}{t} \in S_\Phi + \{v\}.$$

It is easy to see that F has the following properties:

(a) $F(y) > 0$ for any $y \in V \setminus \{0\}$,
(b) $F(\lambda y) = \lambda F(y)$ for any $\lambda > 0$,
(c) $S_F = S_\Phi + \{v\}$.

For any $y \in V \setminus \{0\}$, $F(y)$ can be determined by the following equation,

$$F(y) = \Phi\Big(y - F(y)\, v\Big). \tag{1.8}$$

Moreover, F is a Minkowski norm, i.e., the Hessian $g_{ij} := \frac{1}{2}[F^2]_{y^i y^j}$ is positive definite. The proof is left for the reader. F is called the *Minkowski norm generated by* (Φ, v). One can easily show that if $F = F(y)$ is generated by (Φ, v), then $\Phi = \Phi(y)$ is generated by $(F, -v)$.

Example 1.1.3   For a Euclidean norm $\Phi = |y|$ on V and a vector $v \in V$ with $|v| < 1$, the solution of (1.8) is a Randers norm,

$$F = \frac{\sqrt{(1 - |v|^2)|y|^2 + \langle y, v \rangle^2} - \langle y, v \rangle}{1 - |v|^2}.$$

1.2 Finsler Metrics

We are now ready to introduce Finsler metrics on a manifold. Throughout this book, we always assume that manifolds are C^∞ (smooth), connected and finite dimensional.

Let M be a manifold. For a point $x \in M$, denote by $T_x M$ the *tangent space* of M at x. The *tangent bundle* TM of M is the union of tangent spaces with a natural differential structure,

$$TM := \bigcup_{x \in M} T_x M.$$

Denote the elements in TM by (x, y) where $y \in T_x M$.

Roughly speaking, a Finsler metric on a manifold M is a C^∞ function on the slit tangent bundle $TM_o := TM \setminus \{0\}$, whose restriction to each tangent space $T_x M$ is a Minkowski norm.

Definition 1.2.1 Let M be a manifold. A function $F = F(x, y)$ on TM is called a *Finsler metric* on M if it has the following properties:

(a) $F(x, y)$ is C^∞ on TM_o;
(b) $F_x(y) := F(x, y)$ is a Minkowski norm on $T_x M$ for any $x \in M$.

The pair (M, F) is called a *Finsler manifold*.

A Finsler metric $F = F(x, y)$ on a manifold M is said to be *reversible* if $F(x, -y) = F(x, y)$ for all $y \in T_x M$. We usually do not impose the reversibility condition on Finsler metrics. A Finsler metric F on M is said to be *Riemannian*, if the restriction of F, $F_x(y) := F(x, y)$, is a Euclidean norm on $T_x M$ for any $x \in M$, that is,

$$F_x(y) = \sqrt{\langle y, y \rangle_x}, \qquad y \in T_x M,$$

where $\langle \, , \, \rangle_x$ is an inner product on $T_x M$. We usually denote a Riemannian metric by a family of inner products $g_x = \langle y, y \rangle_x$ on tangent spaces $T_x M$. Clearly, Riemannian metrics are reversible Finsler metrics.

Riemannian metrics are among the most important Finsler metrics. Let us take a look at some special Riemannian metrics.

Example 1.2.2 Let $|\cdot|$ be the standard Euclidean norm on $\mathbb{R}^n$,

$$|y| := \sqrt{\sum_{i=1}^{n} (y^i)^2}.$$

Define $F = F(x, y)$ by

$$F := |y|, \qquad y \in T_x \mathbf{R}^n \cong \mathbf{R}^n.$$

F is a Finsler metric on $\mathbf{R}^n$, which is called the *standard Euclidean metric*.

Example 1.2.3 Let $\mathbf{B}^n \subset (\mathbf{R}^n, |\cdot|)$ be the standard unit ball and let

$$\alpha_{-1} := \frac{\sqrt{|y|^2 - (|x|^2|y|^2 - \langle x, y \rangle^2)}}{1 - |x|^2}, \qquad y \in T_x \mathbf{B}^n \cong \mathbf{R}^n. \qquad (1.9)$$

α_{-1} is a Riemannian metric on $\mathbf{B}^n$, which is called the *Klein metric*. The pair $(\mathbf{B}^n, \alpha_{-1})$ is called the *Klein model*.

Example 1.2.4 Let $\mathbf{S}^n \subset (\mathbf{R}^{n+1}, |\cdot|)$ be the standard unit sphere. For $x \in \mathbf{S}^n$, we identify $T_x \mathbf{S}^n$ with a hypersurface in $\mathbf{R}^{n+1}$ in a natural way. Let

$$\alpha_{+1} := |y|_o, \qquad y \in T_x \mathbf{S}^n \subset \mathbf{R}^{n+1}. \qquad (1.10)$$

Here $|\cdot|_o$ denotes the Euclidean norm on $\mathbf{R}^{n+1}$. Let $\mathbf{S}^n_+$ and $\mathbf{S}^n_-$ denote the upper and lower hemispheres, respectively, and let $\psi_\pm : \mathbf{R}^n \to \mathbf{S}^n_\pm$ be the projection map defined by

$$\psi_\pm(x) := \Big(\frac{x}{\sqrt{1 + |x|^2}}, \ \frac{\pm 1}{\sqrt{1 + |x|^2}} \Big).$$

$\psi_\pm$ sends straight lines in $\mathbf{R}^n$ to great circles on $\mathbf{S}^n_\pm$.

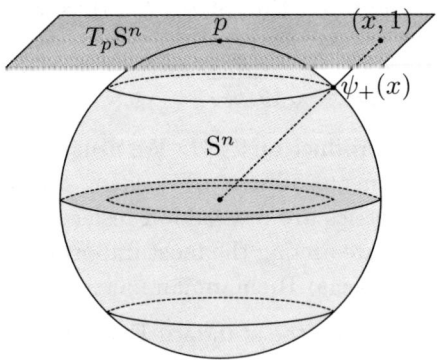

Figure 1.3

The pull-back metric on R^n from S_+^n by ψ_+ is given by

$$\alpha_{+1} = \frac{\sqrt{|y|^2 + (|x|^2|y|^2 - \langle x, y \rangle^2)}}{1 + |x|^2}, \qquad y \in T_x R^n \cong R^n. \qquad (1.11)$$

The pair (R^n, α_{+1}) is called the *projective spherical model*.

The Riemannian metrics in Examples 1.2.2, 1.2.3 and 1.2.4 can be expressed in one single formula.

$$\alpha_\mu := \frac{\sqrt{|y|^2 + \mu(|x|^2|y|^2 - \langle x, y \rangle^2)}}{1 + \mu|x|^2}, \qquad y \in T_x B^n(r_\mu) \cong R^n, \qquad (1.12)$$

where $r_\mu := 1/\sqrt{-\mu}$ if $\mu < 0$ and $r_\mu := +\infty$ if $\mu \geq 0$. The metric α_μ can be expressed as $\alpha_\mu = \sqrt{a_{ij}y^iy^j}$, where

$$a_{ij} = \frac{1}{1 + \mu|x|^2}\Big\{\delta_{ij} - \frac{\mu x^i x^j}{1 + \mu|x|^2}\Big\}.$$

Let $\alpha = \sqrt{a_{ij}(x)y^iy^j}$ be a Riemannian metric and $\beta = b_i(x)y^i$ be a 1-form on an n-dimensional manifold M. Let

$$\|\beta_x\|_\alpha := \sup_{y \in T_x M} \frac{\beta(x, y)}{\alpha(x, y)} = \sqrt{a^{ij}(x)b_i(x)b_j(x)}.$$

Consider the following function

$$F := \alpha\phi(s), \qquad s = \frac{\beta}{\alpha}, \qquad (1.13)$$

where $\phi = \phi(s)$ is a positive C^∞ function on $(-b_o, b_o)$ satisfying

$$\phi(s) - s\phi'(s) + (b^2 - s^2)\phi''(s) > 0, \qquad |s| \leq b < b_o.$$

Then by Lemma 1.1.2, F is a Finsler metric if $\|\beta_x\|_\alpha < b_o$ for any $x \in M$. A Finsler metric in the form (1.13) is called an (α, β)-*metric*.

Let $\phi = 1 + s$. Then $F = \alpha\phi(s)$, where $s = \beta/\alpha$, becomes

$$F = \alpha + \beta.$$

Note that ϕ is positive on $(-1, 1)$ and $\phi(s) - s\phi'(s) + (b^2 - s^2)\phi''(s) = 1$. Thus $F = \alpha + \beta$ is a Finsler metric if and only if $\|\beta_x\|_\alpha < 1$ for any $x \in M$.

One can prove it directly using the formula (1.3) for g_{ij}. The Finsler metric $F = \alpha + \beta$ with $\sup_{x \in M} \|\beta_x\|_\alpha < 1$ is called a *Randers metric* on M.

A typical example of Randers metrics is defined on the ball $B^n(r_\mu) \subset R^n$:

$$F := \frac{\sqrt{|y|^2 + \mu(|x|^2|y|^2 - \langle x, y \rangle^2)} + \sqrt{-\mu}\langle x, y \rangle}{1 + \mu|x|^2}, \qquad (1.14)$$

where $\mu < 0$ and $r_\mu := 1/\sqrt{-\mu}$. The metric when $\mu = -1$ is of particular interest.

$$F := \frac{\sqrt{|y|^2 - (|x|^2|y|^2 - \langle x, y \rangle^2)} + \langle x, y \rangle}{1 - |x|^2}. \qquad (1.15)$$

The metric F in (1.15) is called the *Funk metric* on $B^n(1)$. It has many special geometric properties.

Let $\phi := (1 + s)^2$. $F = \alpha\phi(s)$, where $s = \beta/\alpha$, becomes

$$F = \frac{(\alpha + \beta)^2}{\alpha}.$$

Note that ϕ is positive on $(-1, 1)$ and for any s, b with $|s| \le b < 1$,

$$\phi(s) - s\phi'(s) + (b^2 - s^2)\phi''(s) = 1 - 3s^2 + 2b^2$$
$$> 1 - s^2 + 2(b^2 - s^2) > 0.$$

Thus F is a Finsler metric if and only if $\|\beta_x\|_\alpha < 1$ for any $x \in M$.

A typical example of metrics in the above form is defined on the ball $B^n(r_\mu) \subset R^n$:

$$F = \frac{(\sqrt{|y|^2 + \mu(|x|^2|y|^2 - \langle x, y \rangle^2)} + \sqrt{-\mu}\langle x, y \rangle)^2}{(1 + \mu|x|^2)^2 \sqrt{|y|^2 + \mu(|x|^2|y|^2 - \langle x, y \rangle^2)}}, \qquad (1.16)$$

where $\mu < 0$ and $r_\mu := 1/\sqrt{-\mu}$. The reader should try to find α and β so that $F = (\alpha + \beta)^2/\alpha$. The metric when $\mu = -1$ is of particular interest.

$$F := \frac{(\sqrt{|y|^2 + -(|x|^2|y|^2 - \langle x, y \rangle^2)} + \langle x, y \rangle)^2}{(1 - |x|^2)^2 \sqrt{|y|^2 - (|x|^2|y|^2 - \langle x, y \rangle^2)}}. \qquad (1.17)$$

The metric in (1.17) was constructed by L. Berwald [17]. It has many special geometric properties. We will discuss it later in the book.

One may construct a product Finsler metrics from a pair of Finsler manifolds. Let (M_1, F_1) and (M_2, F_2) be Finsler manifolds. A Finsler metric F on $M := M_1 \times M_2$ is called a *product Finsler metric* of F_1 and F_2 if at any point $x = (x_1, x_2) \in M$,

$$F(x, y) = \begin{cases} F_1(x_1, y_1) & \text{if } y = y_1 \oplus 0 \in T_x M \\ F_2(x_2, y_2) & \text{if } y = 0 \oplus y_2 \in T_x M \end{cases}$$

where $T_x M \cong T_{x_1} M_1 \oplus T_{x_2} M_2$. In this case, (M, F) is called a *product Finsler manifold* of (M_1, F_1) and (M_2, F_2).

For a pair of Finsler manifolds, there is no canonical way to define a product Finsler metrics on the product manifold. When the Finsler metrics are Riemannian, we can define the product Finsler metrics in the following way.

Example 1.2.5 Let α_1 and α_2 be Euclidean norms on vector spaces V_1 and V_2 respectively. Let $f : [0, \infty) \times [0, \infty) \to [0, \infty)$ be a C^∞ function satisfying

$$f(\lambda s, \lambda t) = \lambda f(s, t), \quad \forall \lambda > 0, \quad \text{and} \quad f(s, t) > 0, \quad \forall (s, t) \neq (0, 0). \quad (1.18)$$

Define a function $F : V := V_1 \oplus V_2 \to [0, \infty)$ by

$$F(y) := \sqrt{f\Big([\alpha_1(y_1)]^2, \, [\alpha_2(y_2)]^2\Big)},$$

where $y = y_1 \oplus y_2 \in V = V_1 \oplus V_2$. $F = F(y)$ has the following properties

(a) $F(y) \geq 0$ for any $y \in V$, and $F(y) = 0$ if and only if $y = 0$;
(b) $F(\lambda y) = \lambda F(y)$ for any $y \in V$ and $\lambda > 0$;
(c) $F(y)$ is C^∞ on $V \setminus \{0\}$.

Let $n_1 = \dim V_1$, $n_2 = \dim V_2$ and $n = n_1 + n_2 = \dim V$. We shall assume the following ranges of indices:

$$1 \leq a, b, c \leq n_1, \quad n_1 + 1 \leq \alpha, \beta, \gamma \leq n, \quad 1 \leq i, j, k \leq n.$$

Let $\{\mathbf{b}_a\}$ and $\{\mathbf{b}_\alpha\}$ be bases for V_1 and V_2 respectively. Then $\{\mathbf{b}_i\}$ is a basis for V. Express

$$\alpha_1(y_1) = \sqrt{\bar{g}_{ab} y^a y^b}, \qquad \alpha_2(y_2) = \sqrt{\bar{g}_{\alpha\beta} y^\alpha y^\beta},$$

where $y_1 = y^a \mathbf{b}_a$ and $y_2 = y^\alpha \mathbf{b}_\alpha$. Then $g_{ij} := \frac{1}{2}[F^2]_{y^i y^j}$ are given by

$$(g_{ij}) = \begin{pmatrix} 2f_{ss}\bar{y}_a\bar{y}_b + f_s\bar{g}_{ab} & 2f_{st}\bar{y}_a\bar{y}_\beta \\ 2f_{st}\bar{y}_b\bar{y}_\alpha & 2f_{tt}\bar{y}_\alpha\bar{y}_\beta + f_t\bar{g}_{\alpha\beta} \end{pmatrix}, \qquad (1.19)$$

where $\bar{y}_a := \bar{g}_{ab}y^b$ and $\bar{y}_\alpha := \bar{g}_{\alpha\beta}y^\beta$. By an elementary argument, one can see that (g_{ij}) is positive definite if and only if $f(s,t)$ satisfies the following conditions:

$$f_s > 0, \quad f_t > 0, \quad f_s + 2sf_{ss} > 0, \quad f_t + 2tf_{tt} > 0, \qquad (1.20)$$

and

$$f_s f_t - 2f f_{st} > 0. \qquad (1.21)$$

In this case,

$$\det(g_{ij}) = h\Big([\alpha_1]^2, \ [\alpha_2]^2\Big) \det(\bar{g}_{ab}) \det(\bar{g}_{\alpha\beta}), \qquad (1.22)$$

where

$$h := (f_s)^{n_1-1}(f_t)^{n_2-1}\Big\{ f_s f_t - 2f f_{st} \Big\}.$$

There are lots of functions f satisfying (1.18), (1.20) and (1.21). For example, one can take

$$f(s,t) := \frac{1}{1+\varepsilon}\Big\{ s + t + \varepsilon\big(s^k + t^k\big)^{\frac{1}{k}} \Big\}, \qquad (1.23)$$

where ε is a nonnegative number and k is a positive integer.

By the above construction, one can construct infinitely many product Finsler metrics on $M = M_1 \times M_2$ for any given pair of Riemannian manifolds (M_1, α_1) and (M_2, α_2). Just take a function f as in (1.23) and define

$$F(x,y) := \sqrt{f\Big([\alpha_1(x_1,y_1)]^2, \ [\alpha_2(x_2,y_2)]^2\Big)},$$

where $x = (x_1, x_2) \in M$ and $y = y_1 \oplus y_2 \in T_x M$. Then F is a reversible product Finsler metric on M.

1.3 Length Structure and Volume Form

Every Finsler metric on a manifold defines a *length structure* on piecewise smooth curves. Let M be a C^∞ manifold. A map $c : I = [a, b] \to M$ is called a *piecewise C^∞ curve* if it is continuous and there is a partition $a = t_0 < t_1 < \cdots < t_{n-1} < t_n = b$ such that c restricted to each $[t_{i-1}, t_i]$ is C^∞. For $t \in [t_{i-1}, t_i]$, let $\dot{c}(t) := \frac{dc}{dt}(t) \in T_{c(t)}M$ denote the tangent vector of c. c is said to be *regular* if $\dot{c}(t) \neq 0$, $\forall t \in [t_{i-1}, t_i]$.

Two regular maps $c : I \to M$ and $\bar{c} : \bar{I} \to M$ are said to be *equivalent* if there is a one-to-one and onto piecewise C^∞ map $\varphi : I \to \bar{I}$ such that $\varphi'(t) > 0$ and $\bar{c}(\varphi(t)) = c(t)$, $\forall t \in [t_{i-1}, t_i]$. A (piecewise) C^∞ curve C in a manifold M is an equivalence class of regular (piecewise) C^∞ maps from an interval I into M. For the sake of simplicity, we do not distinguish a regular map $c = c(t)$ and the curve C represented by c.

For a C^∞ curve (represented by) $c : I = [a, b] \to M$, its reverse $c_- : I \to M$ is defined by $c_-(t) := c(b + a - t)$. The class represented by c_- is different from that represented by c. All C^∞ curves in this book are *oriented*.

Consider a piecewise C^∞ curve C from p to q in (M, F). Let C be a piecewise C^∞ curve represented by $c = c(t)$ with $c(a) = p$ and $c(b) = q$. The length of C is defined by

$$\mathcal{L}_F(C) := \int_a^b F\Big(c(t), \dot{c}(t)\Big) dt.$$

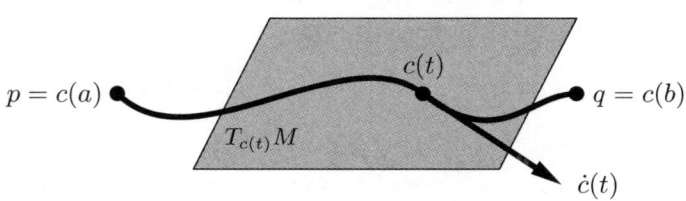

Figure 1.4

If C is represented by another map $\bar{c} = \bar{c}(\bar{t})$ with $\bar{c}(\bar{a}) = p$ and $\bar{c}(\bar{b}) = q$, then there is a positive function $\bar{t} = \varphi(t)$ such that $\bar{c}(\bar{t}) = c(t)$ with $\bar{a} = \varphi(a)$ and $\bar{b} = \varphi(b)$. Then

$$d\bar{t} = \varphi'(t)dt, \qquad \dot{c}(t) = \dot{\bar{c}}(\bar{t})\varphi'(t).$$

We have

$$\int_a^b F\Big(c(t), \dot{c}(t)\Big)dt = \int_a^b F\Big(\bar{c}(\bar{t}), \dot{\bar{c}}(\bar{t})\varphi'(t)\Big)dt$$

$$= \int_a^b F\Big(\bar{c}(\bar{t}), \dot{\bar{c}}(\bar{t})\Big)\varphi'(t)dt = \int_{\bar{a}}^{\bar{b}} F\Big(\bar{c}(\bar{t}), \dot{\bar{c}}(\bar{t})\Big)d\bar{t}.$$

Thus, $\mathcal{L}_F(C)$ is well-defined.

For a pair of points $p, q \in M$, define

$$d_F(p, q) := \inf_C \mathcal{L}_F(C), \tag{1.24}$$

where the infimum is taken over all piecewise C^∞ curves C issuing from p to q. d_F is a function on $M \times M$ with the following properties:

(a) $d_F(p, q) \geq 0$;
(b) $d_F(p, q) = 0$ if and only if $p = q$;
(c) $d_F(p, q) \leq d_F(p, r) + d_F(r, q)$.

The proofs of (a) and (c) are elementary. To prove (b), it suffices to prove the following fact [55]. At every point $x_o \in M$, there is a local coordinate system $\varphi : \mathcal{U} \subset M \to U \subset \mathbf{R}^n$ such that

$$\lambda^{-1}|(y^i)| \leq F(x, y) \leq \lambda|(y^i)|, \qquad y = y^i \frac{\partial}{\partial x^i} \in T_x\mathcal{U},$$

where $\lambda > 1$ is a constant. Choosing a smaller neighborhood $\mathcal{U}$ if necessarily, one can easily show that there is a constant $C > 1$ such that

$$C^{-1}|\varphi(x_2) - \varphi(x_1)| \leq d_F(x_1, x_2) \leq C|\varphi(x_2) - \varphi(x_1)|, \tag{1.25}$$

where $x_1, x_2 \in \mathcal{U}$. This implies (b).

The function d_F is called the *distance function* of F. Let $\Delta \subset M \times M$ denote the diagonal. It can be shown that d_F is C^∞ on $\mathcal{U} \setminus \Delta$ for some neighborhood $\mathcal{U}$ of Δ in $M \times M$. The proof is very technical, so it is omitted here. Conversely, d_F determines the Finsler metric F by

$$F(x, y) = \lim_{\varepsilon \to 0^+} \frac{d_F(x, c(\varepsilon))}{\varepsilon}, \qquad y \in T_xM, \tag{1.26}$$

where $c(t)$ is an arbitrary C^∞ curve with $c(0) = x$ and $\dot{c}(0) = y$.

$$c(\varepsilon)$$

$$c(0) = x$$

Figure 1.5

Clearly, if F is reversible, then the induced distance function d_F is reversible, i.e., $d_F(p, q) = d_F(q, p)$, for all pairs of points $p, q \in M$ and vice versa.

Every Finsler metric $F = F(x, y)$ on an n-dimensional manifold M defines a volume form. At a point $x \in M$, let $\{\mathbf{b}_i\}$ be a basis for $T_x M$ and $\{\theta^i\}$ be the basis for $T_x^* M$ dual to $\{\mathbf{b}_i\}$. Then the following n-form at $x \in M$ is, up to an orientation, well-defined,

$$dV_F := \sigma_F(x)\theta^1 \wedge \cdots \wedge \theta^n,$$

where

$$\sigma_F(x) := \frac{\mathrm{Vol}(\mathrm{B}^n(1))}{\mathrm{Vol}\left((y^i) \in \mathrm{R}^n \mid F(x, y^i \mathbf{b}_i) < 1\right)}. \tag{1.27}$$

Here $\mathrm{Vol}(\cdot)$ denotes the Euclidean volume function on subsets in R^n so that for the unit cubic $\mathcal{U} = [0, 1]^n$, $\mathrm{Vol}(\mathcal{U}) = 1$. The n-form dV_F is called the *Finsler volume form*. For an open subset $\mathcal{U} \subset M$, the volume of $\mathcal{U}$ is defined by

$$\mathrm{Vol}_F(\mathcal{U}) := \int_{\mathcal{U}} dV_F.$$

When F is reversible, the induced distance function d_F is reversible and the Hausdorff measure of d_F can be defined in the usual way. An important fact is that the *Hausdorff measure* of an open subset $\mathcal{U}$ with respect to d_F is equal to $\mathrm{Vol}_F(\mathcal{U})$. See [22].

In general, $\sigma_F(x)$ can't be expressed in terms of elementary functions though $F = F(x, y)$ sometimes is. Nevertheless, $\sigma_F(x)$ is computable for Randers metrics including Riemannian metrics.

First, let us consider a Riemannian metric α on an n-dimensional manifold M. Let

$$\alpha = \sqrt{a_{ij}(x)y^i y^j}, \quad y = y^i \frac{\partial}{\partial x^i}\big|_x \in T_x M.$$

Let A be a matrix such that $A^T A = (a_{ij})$. Then the linear transformation $x = Ay : \mathrm{R}^n \rightarrow \mathrm{R}^n$ sends the convex domain $\mathcal{U}_x := \{(y^i) \in \mathrm{R}^n \mid \sqrt{a_{ij}(x)y^i y^j} < 1\}$ onto the unit ball $\mathrm{B}^n(1)$. We may assume that the Jacobian matrix has positive determinant $\det(A) > 0$. We have

$$\det(A) = \sqrt{\det(a_{ij}(x))}.$$

Observe that

$$\mathrm{Vol}(\mathrm{B}^n(1)) = \int_{\mathrm{B}^n(1)} dx^1 \cdots dx^n$$

$$= \int_{\mathcal{U}_x} \det(A) dy^1 \cdots dy^n = \sqrt{\det(a_{ij}(x))} \mathrm{Vol}(\mathcal{U}_x).$$

That yields,

$$\mathrm{Vol}(\mathcal{U}_x) = \frac{\mathrm{Vol}(\mathrm{B}^n(1))}{\sqrt{\det\left(a_{ij}(x)\right)}}.$$

Then $dV_\alpha = \sigma_\alpha(x) dx^1 \cdots dx^n$ is given by

$$\sigma_\alpha(x) = \sqrt{\det\left(a_{ij}(x)\right)}.$$

Consider a Randers metric $F = \alpha + \beta$, where $\alpha = \sqrt{a_{ij}(x)y^i y^j}$ is a Riemannian metric and $\beta = b_i(x)y^i$ is a 1-form on an n-dimensional manifold M. Let $\Omega_x := \{(y^i) \in \mathrm{R}^n \mid F(x, y^i \frac{\partial}{\partial x^i}\big|_x) < 1\}$. By an elementary argument using linear algebra, we obtain

$$\mathrm{Vol}(\Omega_x) = \frac{\mathrm{Vol}(\mathrm{B}^n(1))}{\left(1 - \|\beta_x\|_\alpha^2\right)^{(n+1)/2} \sqrt{\det\left(a_{ij}(x)\right)}},$$

where $\|\beta_x\|_\alpha$ denotes the norm of β at x with respect to α_x. Plugging the above formula into (1.27) yields

$$\sigma_F(x) = \left(1 - \|\beta_x\|_\alpha^2\right)^{(n+1)/2} \sigma_\alpha(x).$$

Thus

$$dV_F = \left(1 - \|\beta_x\|_\alpha^2\right)^{(n+1)/2} dV_\alpha. \tag{1.28}$$

Note that for any open subset $\Omega \subset M$,

$$\int_\Omega dV_F \le \int_\Omega dV_\alpha.$$

Equality holds if and only if $F = \alpha$.

Example 1.3.1 Consider the Randers metric $F = \alpha + \beta$ on the unit ball $B^n(1) \subset \mathrm{R}^n$, where $\alpha = \sqrt{a_{ij}(x)y^i y^j}$ and $\beta = b_i(x)y^i$ are given by

$$\alpha := \frac{\sqrt{|y|^2 - (|x|^2|y|^2 - \langle x, y \rangle^2)}}{1 - |x|^2},$$

$$\beta := \frac{\langle x, y \rangle}{1 - |x|^2} + \frac{\langle a, y \rangle}{1 + \langle a, x \rangle},$$

where $y \in T_x\mathrm{R}^n \cong \mathrm{R}^n$, $a \in B^n(1)$, $|\cdot|$ and $\langle\ ,\ \rangle$ denote the standard Euclidean norm and inner product in R^n, respectively. When $a = 0$, F is the Funk metric defined in (1.15).

Applying Lemma 1.1.1 to the following matrix

$$a_{ij} = \frac{1}{1 - |x|^2}\left\{\delta_{ij} + \frac{x^i x^j}{1 - |x|^2}\right\},$$

we obtain

$$\det(a_{ij}) = \frac{1}{(1 - |x|^2)^{n+1}}.$$

Then

$$dV_\alpha = \left(1 - |x|^2\right)^{-\frac{n+1}{2}} dx^1 \cdots dx^n.$$

By Lemma 1.1.1,

$$a^{ij} = (1 - |x|^2)\left\{\delta^{ij} - x^i x^j\right\}.$$

Then the norm of β is given by

$$\|\beta_x\|_\alpha = \sqrt{a^{ij}(x)b_i(x)b_j(x)} = 1 - \frac{(1 - |x|^2)(1 - |a|^2)}{(1 + \langle a, x \rangle)^2}. \tag{1.29}$$

Plugging the above formulas into (1.28) yields

$$dV_F = \left[\frac{1 - |a|^2}{(1 + \langle a, x \rangle)^2} \right]^{\frac{n+1}{2}} dx^1 \cdots dx^n.$$

1.4 Navigation Problem

In this section, we will discuss Randers metrics from a navigation point of view. We shall see that non-Riemannian metrics are not avoidable even though we live in a Riemannian world.

Consider an object moving in a metric space, such as Euclidean space, pushed by an interval force and an external force field. The shortest time problem is to determine a curve from one point to another in the space, along which it takes the least time for the object to travel. This problem in some special cases was studied by E. Zermelo [102], hence called the *Zermelo navigation problem*. Here we shall discuss the navigation problem in the most general case. Suppose that an object on a Finsler manifold (M, Φ) is pushed by an internal force U with constant length, $\Phi(x, U_x) = c$, and while it is pushed by an external force field V with $\Phi(x, -V_x) < c$. The combined force at x is $T_x := U_x + V_x$. The condition, $\Phi(x, -V_x) < c$, guarantees that the object can move forward in any direction.

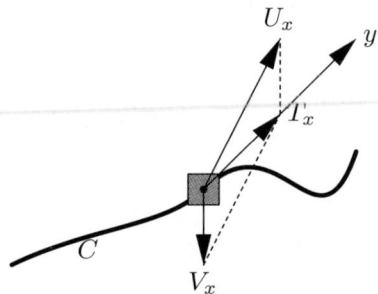

Figure 1.6

Due to the friction, the object moves on M at a speed proportional to the combined force T. For the sake of simplicity, one may assume that $c = 1$ and the velocity vector at any point $x \in M$ is equal to T_x. Given a pair of points $p, q \in M$, let C be an arbitrary piecewise C^∞ curve in M. Since $\Phi(x, U_x) = 1$, we have

$$\Phi\left(x, \ T_x - V_x\right) = \Phi(x, U_x) = 1. \tag{1.30}$$

On the other hand, for any vector $y \in T_x M \setminus \{0\}$, there is a unique solution $F = F(x, y) > 0$ to the following equation

$$\Phi\left(x, \ \frac{y}{F} - V_x\right) = 1. \tag{1.31}$$

Observe that for any $\lambda > 0$,

$$1 = \Phi\left(x, \ \frac{\lambda y}{\lambda F(x, y)} - V_x\right) = \Phi\left(x, \ \frac{\lambda y}{F(x, \lambda y)} - V_x\right).$$

By the uniqueness,

$$F(x, \lambda y) = \lambda F(x, y).$$

One can show that $F_x := F|_{T_x M}$ is a Minkowski norm on $T_x M$. Thus $F = F(x, y)$ is a Finsler metric on M. Comparing (1.30) and (1.31), one can see that the combined force T_x has unit F-length,

$$F(x, T_x) = 1. \tag{1.32}$$

This observation leads to the following

Lemma 1.4.1 *Let (M, Φ) be a Finsler manifold and V be a vector field on M with $\Phi(x, -V_x) < 1, \ \forall x \in M$. Define $F : TM \to [0, \infty)$ by (1.31). For any piecewise C^∞ curve C in M, the F-length of C is equal to the time for which the object travels along it.*

Proof: Let $c : [0, t_o] \to M$ be the parametrization of C such that the velocity vector $\dot{c}(t) = T_{c(t)}$. Then t_o is the time for which the object travels along C. It follows from (1.32) that

$$F\left(c(t), \dot{c}(t)\right) = 1.$$

This implies

$$t_o = \int_0^{t_o} F\left(c(t), \dot{c}(t)\right) dt = \mathcal{L}_F(C).$$

Q.E.D.

For a pair $\{\Phi, V\}$ on a manifold M, where $\Phi = \Phi(x, y)$ is a Finsler metric and V is a vector field with $\Phi(x, -V_x) < 1$, we define a Finsler metric $F = F(x, y)$ by (1.31). The Finsler metric F can also be defined in the following way. First, define Φ^* and V^* on T^*M by

$$\Phi^*(x, \xi) := \sup_{y \in T_x M} \frac{\eta(y)}{\Phi(x, y)}, \qquad V^*(\xi) := \xi(V_x), \qquad \xi \in T_x^* M.$$

Then $F^* := \Phi^* + V^*$ is a co-Finsler metric on M and F is dual to F^*, i.e.,

$$F(x, y) = \sup_{\xi \in T_x^* M} \frac{\eta(y)}{F^*(x, \xi)}.$$

The proof is left to the reader.

Lemma 1.4.2 *Let $\Phi = \Phi(x, y)$ be a Finsler metric on an n-dimensional manifold M and $V = V^i(x)\frac{\partial}{\partial x^i}$ be an arbitrary vector field on M with $\Phi(x, -V_x) < 1$, $x \in M$. Let $F = F(x, y)$ denote the Finsler metric on M defined by (1.31). Then the Finsler volume forms of F and Φ are equal,*

$$dV_F = dV_\Phi. \tag{1.33}$$

Proof. Fix a basis $\{\mathbf{b}_i\}$ for $T_x M$ and let $V_x := v^i \mathbf{b}_i$. Let

$$\mathcal{U}_\Phi := \left\{ (y^i) \in \mathbf{R}^n \mid \Phi(x, y^i \mathbf{b}_i) < 1 \right\},$$

$$\mathcal{U}_F := \left\{ (y^i) \in \mathbf{R}^n \mid F(x, y^i \mathbf{b}_i) < 1 \right\}.$$

From the definition of F, we have

$$\mathcal{U}_F = \mathcal{U}_\Phi + (v^i).$$

Since shifting does not change the Euclidean volume, $\mathrm{Vol}(\mathcal{U}_\Phi) = \mathrm{Vol}(\mathcal{U}_F)$. This implies that

$$\sigma_F(x) = \sigma_\Phi(x).$$

Thus $dV_F = dV_\Phi$. Q.E.D.

The above proposition shows that the volume of an open subset on a Finsler manifold is not disturbed by any vector field.

Example 1.4.3 Let $\phi = \phi(y)$ be a Minkowski norm on $\mathbf{R}^n$ and

$$\mathcal{U} := \left\{ y \in \mathbf{R}^n \mid \phi(y) < 1 \right\}.$$

Let $\Phi(x, y) := \phi(y)$, where $y \in T_x V \cong V$, and $V_x := -x$, where $x \in V$. Φ is a Minkowski metric on $\mathbf{R}^n$ and V is a radial vector field toward the origin. Observe that

$$\Phi(x, -V_x) = \phi(x) < 1, \qquad x \in \mathcal{U}.$$

For a non-zero vector $y \in T_x \mathcal{U} \setminus \{0\} \cong \mathbf{R}^n \setminus \{0\}$, define $\Theta = \Theta(x, y) > 0$ by

$$\Phi\left(x, \ \frac{y}{\Theta(x, y)} - V_x\right) = 1. \tag{1.34}$$

Then $\Theta = \Theta(x, y)$ is a Finsler metric on $\mathcal{U}$, which is called the *Funk metric* on $\mathcal{U}$. The Funk metric $\Theta = \Theta(x, y)$ can also be defined by

$$z := x + \frac{y}{\Theta(x, y)} \in \partial \mathcal{U}.$$

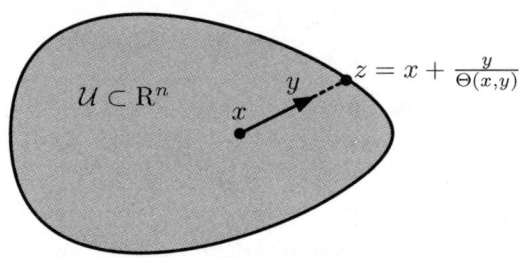

Figure 1.7

Equation (1.34) can be written as

$$\Theta(x, y) = \phi\Big(y + \Theta(x, y)x\Big). \tag{1.35}$$

Differentiating (1.35) with respect to x^k and y^k respectively, one obtains

$$\Big(1 - \phi_{w^l}(w)x^l\Big)\Theta_{x^k}(x, y) = \phi_{w^k}(w)\Theta(x, y), \tag{1.36}$$

$$\Big(1 - \phi_{w^l}(w)x^l\Big)\Theta_{y^k}(x, y) = \phi_{w^k}(w), \tag{1.37}$$

where $w := y + \Theta(x, y)x$. It follows from (1.36) and (1.37) that

$$\Theta_{x^k} = \Theta\Theta_{y^k}. \tag{1.38}$$

The above argument is given by T. Okada [77].

If $\phi = |y|$ is the standard Euclidean norm on $\mathbf{R}^n$, $\mathcal{U} = B^n(1)$ is the unit ball in $\mathbf{R}^n$. In this case, $\Theta = F$ as defined in (1.15).

Given a Riemannian metric $h = \sqrt{h_{ij}(x)y^iy^j}$ and a vector field $V = V^i(x)\frac{\partial}{\partial x^i}$ on a manifold M with $h(x, -V_x) = \sqrt{h_{ij}(x)V^i(x)V^j(x)} < 1$, one can define a Finsler $F = F(x,y)$ by (1.31), i.e.,

$$h\left(x, \frac{y}{F} - V\right) = \sqrt{h_{ij}\left(\frac{y^i}{F} - V^i\right)\left(\frac{y^j}{F} - V^j\right)} = 1. \qquad (1.39)$$

Solving (1.39) for F, one obtains $F = \alpha + \beta$, where $\alpha = \sqrt{a_{ij}(x)y^iy^j}$ and $\beta = b_i(x)y^i$ are given by

$$a_{ij} = \frac{(1 - h_{pq}V^pV^q)h_{ij} + h_{ip}h_{jq}V^pV^q}{(1 - h_{pq}V^pV^q)^2}, \qquad (1.40)$$

$$b_i = -\frac{h_{ip}V^p}{1 - h_{pq}V^pV^q}. \qquad (1.41)$$

It is easy to show that

$$\|\beta_x\|_\alpha = \sqrt{a^{ij}b_ib_j} = \sqrt{h_{ij}V^iV^j} = h(x, -V_x) < 1. \qquad (1.42)$$

Thus F is a Randers metric.

Conversely, every Randers metric $F = \alpha + \beta$ on a manifold M can be constructed from a Riemannian metric h and a vector field V on M. The construction is given as follows. Let $\alpha = \sqrt{a_{ij}y^iy^j}$ and $\beta = b_iy^i$. Define

$$h_{ij} := (1 - \|\beta_x\|^2)\left\{a_{ij} - b_ib_j\right\}, \qquad (1.43)$$

$$V^i := -\frac{a^{ij}b_j}{1 - \|\beta_x\|_\alpha^2}. \qquad (1.44)$$

Then F is given by (1.39) for $h = \sqrt{h_{ij}(x)y^iy^j}$ and $V = V^i(x)\frac{\partial}{\partial x^i}$. Moreover, (1.42) holds. Thus $h(x, -V_x) < 1$ for $x \in M$. See [45] and [46] for a similar type of duality between Randers metrics defined as a function on TM and Randers co-metrics defined as a function on T^*M.

Example 1.4.4 Let $B^n \subset \mathbf{R}^n$ be the standard unit ball and let

$$h := \frac{\sqrt{1 - |a|^2}}{1 + \langle a, x\rangle}\sqrt{|y|^2 - \frac{2\langle a, y\rangle\langle x, y\rangle}{1 + \langle a, x\rangle} - \frac{(1 - |x|^2)\langle a, y\rangle^2}{(1 + \langle a, x\rangle)^2}}, \qquad (1.45)$$

$$V := -\frac{1 + \langle a, x \rangle}{1 - |a|^2}(x + a), \tag{1.46}$$

where $y \in T_x B^n = R^n$ and $a \in R^n$ is a constant vector with $|a| < 1$. By (1.39), one obtains

$$F = \frac{\sqrt{|y|^2 - (|x|^2|y|^2 - \langle x, y \rangle^2)}}{1 - |x|^2} + \frac{\langle x, y \rangle}{1 - |x|^2} + \frac{\langle a, y \rangle}{1 + \langle a, x \rangle}. \tag{1.47}$$

Both $h = h(x, y)$ and $F = F(x, y)$ have some special geometric properties. See Examples 3.4.2 and 3.4.6 below. When $a = 0$, F is the Funk metric on B^n defined in (1.15).

1.5 Cartan Torsion

To characterize Euclidean norms among Minkowski norms, E. Cartan introduces a quantity for Minkowski norms [23].

Let $F = F(y)$ be a Minkowski norm on a vector space V. For a vector $y \in V \setminus \{0\}$, let

$$\mathbf{C}_y(u, v, w) := \frac{1}{4} \frac{\partial^3}{\partial s \partial t \partial r} \left[F^2(y + su + tv + rw) \right]_{s=t=r=0},$$

where $u, v, w \in V$. Each $\mathbf{C}_y$ is a symmetric trilinear form on V. We call the family $\mathbf{C} := \{\mathbf{C}_y \mid y \in V \setminus \{0\}\}$ the *Cartan torsion*.

Let $\{\mathbf{b}_i\}$ be a basis for V. Let $g_{ij} := \mathbf{g}_y(\mathbf{b}_i, \mathbf{b}_j)$, $C_{ijk} := \mathbf{C}_y(\mathbf{b}_i, \mathbf{b}_j, \mathbf{b}_k)$. Then

$$g_{ij} = \frac{1}{2}[F^2]_{y^i y^j},$$

$$C_{ijk} = \frac{1}{4}[F^2]_{y^i y^j y^k} = \frac{1}{2}\frac{\partial}{\partial y^k}(g_{ij}).$$

Define the mean value of the Cartan torsion by

$$\mathbf{I}_y(u) := \sum_{i=1}^n g^{ij}(y)\mathbf{C}_y(u, \mathbf{b}_i, \mathbf{b}_j), \qquad u \in V.$$

We call the family $\mathbf{I} := \{\mathbf{I}_y \mid y \in V \setminus \{0\}\}$ the *mean Cartan torsion*. Observe that

$$\frac{\partial}{\partial y^i}\left[\det(g_{jk}) \right] = \det(g^{jk})g^{pq}\frac{\partial g_{pq}}{\partial y^i} = 2\det(g^{jk})g^{pq}C_{ipq}.$$

We have

$$I_i = g^{jk} C_{ijk} = \frac{\partial}{\partial y^i} \left[\ln \sqrt{\det \left(g_{jk} \right)} \right]. \tag{1.48}$$

It follows from the homogeneity of F that

$$\mathbf{C}_y(y, v, w) = \mathbf{C}_y(u, y, w) = \mathbf{C}_y(u, v, y) = 0 \tag{1.49}$$

and

$$\mathbf{I}_y(y) = 0. \tag{1.50}$$

Moreover,

$$\mathbf{C}_{\lambda y} = \lambda^{-1} \mathbf{C}_y, \qquad \mathbf{I}_{\lambda y} = \lambda^{-1} \mathbf{I}_y, \qquad \lambda > 0. \tag{1.51}$$

From (1.49)-(1.51), one can see that $\mathbf{C}_y$ and $\mathbf{I}_y$ depend only on the geometry of the indicatrix S_F of F. Intuitively, the indicatrix of F can be viewed as a color pattern on V, then $\mathbf{C}_y$ (resp. $\mathbf{I}_y$) is the rate (resp. average rate) of tangential change of the color pattern at y.

It is obvious that F is Euclidean if and only if $\mathbf{C}_y = 0$ for any $y \in V \backslash \{0\}$. In fact, Euclidean norms can be characterized by the mean Cartan torsion. The following result is due to Deicke [34].

Theorem 1.5.1 ([34]) *A Minkowski norm on a vector space* V *is Euclidean if and only if* $\mathbf{I} = 0$.

The proof does not fit in this book, so it is omitted. One can see [5] for a proof.

To characterize Randers norms among Minkowski norms, M. Matsumoto introduces the following quantity [64] [66]. For $y = y^i \mathbf{b}_i \in V$, define

$$M_{ijk} := C_{ijk} - \frac{1}{n+1} \left\{ I_i h_{jk} + I_j h_{ik} + I_k h_{ij} \right\}, \tag{1.52}$$

where $h_{ij} := FF_{y^i y^j} = g_{ij} - \frac{1}{F^2} g_{ip} y^p g_{jq} y^q$. Let

$$\mathbf{M}_y(u, v, w) := M_{ijk}(x, y) u^i v^j w^k, \tag{1.53}$$

where $u = u^i \mathbf{b}_i$, $v = v^j \mathbf{b}_j$ and $w = w^k \mathbf{b}_k$. Each $\mathbf{M}_y$ is a symmetric trilinear form on V. We call the family $\mathbf{M} := \{\mathbf{M}_y \mid y \in V \setminus \{0\}\}$ the *Matsumoto torsion*. Clearly, $\mathbf{M} = 0$ for all two-dimensional Minkowski norms.

Example 1.5.2 ([64]) Let $F = \alpha + \beta$ be a Randers norm on a vector space V, where $\alpha = \sqrt{a_{ij}y^i y^j}$ and $\beta = b_i y^i$ with $\|\beta\|_\alpha < 1$. Then $g_{ij} := \frac{1}{2}[F^2]_{y^i y^j}(y)$ are given by (1.3) and $\det(g_{ij})$ is given by (1.4). Note that $\det(a_{ij})$ is independent of y. By (1.48), one obtains

$$I_i = \frac{\partial}{\partial y^i} \ln \sqrt{\left(\frac{\alpha + \beta}{\alpha}\right)^{n+1} \det(a_{ij})}$$

$$= \frac{n+1}{2(\alpha + \beta)} \cdot \left(b_i - \frac{y_i}{\alpha}\frac{\beta}{\alpha}\right). \tag{1.54}$$

Differentiating (1.3) with respect to y^k yields

$$C_{ijk} = \frac{1}{n+1}\left\{I_i h_{jk} + I_j h_{ik} + I_k h_{ij}\right\}, \tag{1.55}$$

where $h_{ij} := F F_{y^i y^j}$ are given by

$$h_{ij} = \frac{\alpha + \beta}{\alpha}\left(a_{ij} - \frac{y_i y_j}{\alpha^2}\right).$$

This implies that $M_{ijk} = 0$.

Minkowski norms with $\mathbf{M} = 0$ are said to be *C-reducible*. It turns out that every C-reducible Minkowski norm is a Randers norm in dimension $n \geq 3$.

Proposition 1.5.3 ([64], [70]) *Let F be a Minkowski norm on a vector space V of dimension $n \geq 3$. The Matsumoto torsion $\mathbf{M} = 0$ if and only if F is a Randers norm.*

The proof does not fit in this book, and so is omitted. See [70] for more details.

Given a Minkowski space (V, F), using the family of inner products $\mathbf{g}_y$ on V, one can define the norm of $\mathbf{I}, \mathbf{C}$ and $\mathbf{M}$ in a natural way.

$$\|\mathbf{I}\| := \sup_{y,u \in V \setminus \{0\}} \frac{F(y)|\mathbf{I}_y(u)|}{\sqrt{\mathbf{g}_y(u,u)}},$$

$$\|\mathbf{C}\| := \sup_{y,u,v,w \in V \setminus \{0\}} \frac{F(y)|\mathbf{C}_y(u,v,w)|}{\sqrt{\mathbf{g}_y(u,u)\mathbf{g}_y(v,v)\mathbf{g}_y(w,w)}},$$

$$\|\mathbf{M}\| := \sup_{y,u,v,w \in V \setminus \{0\}} \frac{F(y)|\mathbf{M}_y(u,v,w)|}{\sqrt{\mathbf{g}_y(u,u)\mathbf{g}_y(v,v)\mathbf{g}_y(w,w)}}.$$

By (1.52), $\|\mathbf{M}\|$ is bounded by $\|\mathbf{C}\|$. It is easy to construct a family of Minkowski norms F_i on $\mathbf{R}^n$ with $\|\mathbf{C}_i\| \to +\infty$ as $i \to +\infty$, where $\mathbf{C}_i$ denotes the Cartan torsion of F_i.

The (mean) Cartan torsion of any Randers norm is bounded from above by a number depending only on the dimension.

Lemma 1.5.4   ([49]) *Let $F = \alpha + \beta$ be a Randers norm on an n-dimensional vector space $\mathbf{V}$. Then*

$$\|\mathbf{I}\| = \frac{n+1}{\sqrt{2}} \sqrt{1 - \sqrt{1 - \|\beta\|_\alpha^2}} < \frac{n+1}{\sqrt{2}}. \tag{1.56}$$

Proof. Let $\alpha = \sqrt{a_{ij} y^i y^j}$ and $\beta = b_i y^i$. Then $g_{ij} := \frac{1}{2}[F^2]_{y^i y^j}$ are given by (1.3). By (1.3) and Lemma 1.1.1, one can find the inverse matrix $(g^{ij}) = (g_{ij})^{-1}$.

$$g^{ij} = \frac{\alpha}{F} a^{ij} - \frac{\alpha}{F^2}(b^i y^j + b^j y^i) + \frac{\alpha\|\beta\|_\alpha^2 + \beta}{\alpha^3} y^i y^j. \tag{1.57}$$

By (1.54) and (1.57), one obtains

$$I_i I_j g^{ij} = \left(\frac{n+1}{2F(y)}\right)^2 \frac{\alpha(y)}{F(y)} \left\{ \|\beta\|_\alpha^2 - \left(\frac{\beta(y)}{\alpha(y)}\right)^2 \right\}. \tag{1.58}$$

Since $|\beta(y)| \le \|\beta\|_\alpha \alpha(y)$, we can write $\beta(y) = \|\beta\|_\alpha \alpha(y) \cos\theta$, where $0 \le \theta \le 2\pi$. Assume that y is a unit vector, i.e., $F(y) = \alpha(y) + \beta(y) = 1$. Then

$$\alpha(y) = 1 - \beta(y) = 1 - \|\beta\|_\alpha \alpha(y) \cos\theta.$$

Thus

$$\alpha(y) = \frac{1}{1 + \|\beta\|_\alpha \cos\theta}.$$

Plugging it into (1.58) yields

$$I_i I_j g^{ij} = \left(\frac{n+1}{2}\right)^2 \frac{\|\beta\|_\alpha^2 \sin^2\theta}{1 + \|\beta\|_\alpha \cos\theta},$$

$$\|\mathbf{I}\|^2 = \frac{(n+1)^2}{2}\left(1 - \sqrt{1 - \|\beta\|_\alpha^2}\right).$$

This gives the upper bound (1.56) immediately.   Q.E.D.

It follows from (1.55) and (1.56) that

$$\|\mathbf{C}\| \le \frac{3}{\sqrt{2}}\sqrt{1 - \sqrt{1 - \|\beta\|_\alpha^2}} < \frac{3}{\sqrt{2}}. \tag{1.59}$$

Namely, the Cartan torsion is uniformly bounded by $3/\sqrt{2}$. The bound (1.59) for two-dimensional Randers norms is given in Exercise 11.2.6 in [5] which is suggested by Brad Lackey.

Example 1.5.5 Consider the generalized Funk metric $F = \alpha + \beta$ on the unit ball $B^n(1) \subset R^n$,

$$F = \frac{\sqrt{|y|^2 - (|x|^2|y|^2 - \langle x, y \rangle^2)}}{1 - |x|^2} + \frac{\langle x, y \rangle}{1 - |x|^2} + \frac{\langle a, y \rangle}{1 + \langle a, x \rangle}.$$

Let $\|\mathbf{I}\|_x$ denote the norm of the mean Cartan torsion at $x \in B^n(1)$. By (1.29) and (1.56), one obtains

$$\|\mathbf{I}\|_x = \frac{n+1}{\sqrt{2}}\left\{1 - \frac{\sqrt{(1 - |x|^2)(1 - |a|^2)}}{1 + \langle a, x \rangle}\right\}.$$

Note that at $x = -a$, $\mathbf{I}_x = 0$, namely, F_x is Euclidean. However, as $x \to \partial B^n(1)$, $\|\mathbf{I}\|_x \to (n+1)/\sqrt{2}$. The point $x = -a$ can be regarded as the *Euclidean center* of F.

Chapter 2

Structure Equations

In 1943, the first author introduced a connection for Finsler metrics and gave a solution of the local congruence, i.e., a complete system of local invariants which ensures that two Finsler structures differ by a change of coordinates [28]. This connection is a natural generalization of the Levi-Civita connection in the Riemannian case and seems to be the right analytical basis of the subject. This connection is now called the Chern connection. The aim of this chapter is to give a short derivation of the Chern connection, introduce various notions of curvatures and derive important relationships among these quantities (see also [4], [29], [30], [31]).

2.1 Chern Connection

The Chern connection on a Finsler manifold is a linear connection on the pull-back tangent bundle. Before we introduce the Chern connection, let us give a brief description of vector bundles and linear connections on a vector bundle.

A k-dimensional *vector bundle* over a C^∞ manifold N is a C^∞ manifold $\mathcal{V}$ with an *onto* C^∞ map $\pi : \mathcal{V} \to N$ such that for any coordinate domain $\mathcal{U} \subset N$, $\pi^{-1}(\mathcal{U})$ is diffeomorphic to $\mathcal{U} \times \mathrm{R}^k$ such that $\pi^{-1}(x)$ is diffeomorphic to $\{x\} \times \mathrm{R}^k$ for any $x \in \mathcal{U}$ by the restriction of the diffeomorphism. The set $\mathcal{V}_x := \pi^{-1}(x)$ is called the *fiber* at x. We usually denote a vector bundle $\pi : \mathcal{V} \to N$ by $\mathcal{V}$.

For a vector bundle $\mathcal{V}$ over a manifold N, a section of $\mathcal{V}$ is a map $X : N \to \mathcal{V}$ such that $X(x) \in \mathcal{V}_x$ for any $x \in N$. A local frame of $\mathcal{V}$ is a set of local C^∞ sections $\{\mathbf{e}_i\}_{i=1}^k$ of $\mathcal{V}$ defined on some open subset $\mathcal{U} \subset M$

such that for any $x \in \mathcal{U}$, the set $\{\mathbf{e}_i(x)\}_{i=1}^{k}$ is a basis for the fiber $\mathcal{V}_x$ at x. Given a local frame $\{\mathbf{e}_i\}$ of $\mathcal{V}$, any section X of $\mathcal{V}$ can be locally expressed by $X = X^i \mathbf{e}_i$. Then X is C^∞ if and only if all coefficients X^i are C^∞. We shall denote by $C^\infty(\mathcal{V})$ the space of all C^∞ sections of $\mathcal{V}$.

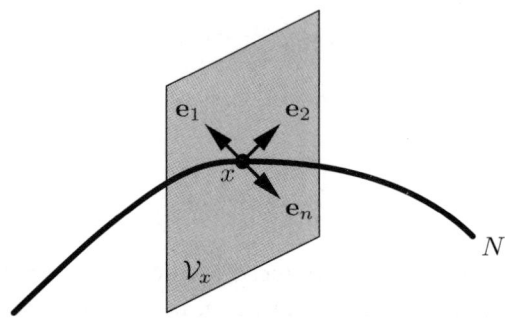

Figure 2.1

One can view a vector bundle $\mathcal{V}$ over a manifold N as a union of vector spaces $\mathcal{V}_x$ indexed on N, $\mathcal{V} = \bigcup_{x \in N} \mathcal{V}_x$. Let $\mathcal{V}_x^*$ denote the dual vector space of $\mathcal{V}_x$. By definition, $\mathcal{V}_x^*$ is the vector space of linear functionals on $\mathcal{V}_x$. Then $\mathcal{V}^* := \bigcup_{x \in N} \mathcal{V}_x^*$ is a vector bundle over N. We call $\mathcal{V}^*$ the *dual vector bundle* of $\mathcal{V}$.

Let $\mathcal{V}$ be a vector bundle over a manifold N. A *linear connection* ∇ on $\mathcal{V}$ is a family of *linear* maps $\nabla : T_x N \times C^\infty(\mathcal{V}) \to \mathcal{V}_x$, i.e.,

$$\nabla : (v, X) \in T_x N \times C^\infty(\mathcal{V}) \to \nabla_v X \in \mathcal{V}_x$$

with the following additional condition:

$$\nabla_v(fX) = df(v)X + f(x)\nabla_v X, \quad f \in C^\infty(M).$$

For any local frame $\{\mathbf{e}_i\}$ for $\mathcal{V}$, let $X = X^i \mathbf{e}_i$. Then

$$\nabla_v X = \left\{ dX^i(v) + X^j \omega_j{}^i(v) \right\} \mathbf{e}_i,$$

where $\{\omega_j{}^i\}$ is a set of local 1-forms on N. $\{\omega_j{}^i\}$ are called the *connection forms* of ∇ with respect to $\{\mathbf{e}_i\}_{i=1}^n$. By removing v in the above identity, we can express $\nabla X : T_x N \to \mathcal{V}_x$ or $\nabla X \in T_x^* N \otimes \mathcal{V}_x$ as follows,

$$\nabla X = \left\{ dX^i + X^j \omega_j{}^i \right\} \otimes \mathbf{e}_i, \quad X = X^i \mathbf{e}_i.$$

Set

$$\Omega_j{}^i := d\omega_j{}^i - \omega_j{}^k \wedge \omega_k{}^i.$$

Each $\Omega_j{}^i$ is a local 2-form. $\{\Omega_j{}^i\}$ are called the *curvature forms* of ∇ with respect to $\{\mathbf{e}_i\}$. Let $\{\omega^i\}$ denote the dual basis of $\{\mathbf{e}_i\}$, then $\Omega :=$ $\Omega_j{}^i\,\omega^j \otimes \mathbf{e}_i$ is a well-defined tensor over N, which is a C^∞ section of $T^*N \otimes \mathcal{V}$.

Let M be a connected C^∞ manifold. Let

$$TM_o := TM \setminus \{0\} = \Big\{ y \in T_xM \ \Big|\ y \neq 0,\ x \in M \Big\}.$$

TM_o is called the *slit tangent bundle* over M. The natural projection $\pi :$ $TM_o \to M$ pulls back TM to a vector bundle π^*TM over TM_o. The fiber at a point $(x, y) \in TM_o$ is defined by

$$\pi^*TM|_{(x,y)} := \Big\{ (x, y, v) \ \Big|\ v \in T_xM \Big\} \cong T_xM.$$

In other words, $\pi^*TM|_{(x,y)}$ is just a copy of T_xM. π^*TM is called the *pull-back tangent bundle*. Similarly, we define the pull-back cotangent bundle π^*T^*M whose fiber at (x, y) is a copy of T_x^*M. Therefore π^*T^*M can be viewed as the dual vector bundle of π^*TM by setting

$$(x, y, \theta)(x, y, v) = \theta(v), \qquad (\theta, v) \in T_x^*M \times T_xM.$$

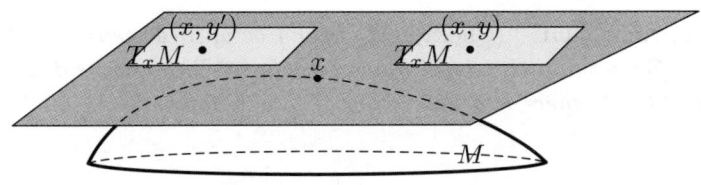

Figure 2.2

Take a standard local coordinate system (x^i, y^i) in TM, where (x^i) is a local coordinate system in M and y^i's are coefficients of $y = y^i \frac{\partial}{\partial x^i}|_x$. Let $\{\frac{\partial}{\partial x^i}, \frac{\partial}{\partial y^i}\}$ and $\{dx^i, dy^i\}$ be the natural local frame and coframe for $T(TM_o)$ and $T^*(TM_o)$, respectively. Then $VTM := \operatorname{span}\{\frac{\partial}{\partial y^i}\}$ is a well-defined subbundle of $T(TM_o)$, which is called the *vertical tangent bundle* of

M. π^*T^*M can be naturally identified with the *horizontal tangent cotangent bundle*, $HT^*M := \operatorname{span}\{dx^i\}$ of $T^*(TM_o)$. Thus HT^*M and π^*T^*M can be viewed as the dual vector bundle of π^*TM. Let

$$\partial_i := \left(x, y \frac{\partial}{\partial x^i}|_x\right).$$

Then $\{\partial_i\}$ is a local frame for π^*TM.

The vector bundle π^*TM has a canonical section $\mathcal{Y}$ defined by

$$\mathcal{Y}_{(x,y)} := (x, y, y).$$

At $y = y^i \frac{\partial}{\partial x^i}|_x \in T_x M$, $\mathcal{Y}$ can be expressed as

$$\mathcal{Y} = y^i \partial_i. \tag{2.1}$$

Let F be a Finsler metric on M and let

$$g_{ij} := \frac{1}{2}[F^2]_{y^i y^j}(x, y), \qquad C_{ijk} := \frac{1}{4}[F^2]_{y^i y^j y^k}(x, y).$$

Define

$$\mathcal{G} := g_{ij}\, dx^i \otimes dx^j, \qquad \mathcal{C} := C_{ijk}\, dx^i \otimes dx^j \otimes dx^k. \tag{2.2}$$

$\mathcal{G}$ and $\mathcal{C}$ are tensors on $TM_o := TM \setminus \{0\}$. $\mathcal{G}$ and $\mathcal{C}$ are called the *fundamental tensor* and the *Cartan tensor* respectively. With these tensors, we can state the following

Theorem 2.1.1   (Chern) *Let (M, F) be an n-dimensional Finsler manifold. For an arbitrary local frame $\{\mathbf{e}_i\}$ for π^*TM and its dual coframe $\{\omega^i\}$ for π^*T^*M, there is a unique set of local 1-forms $\{\omega_j{}^i\}$ on $TM \setminus \{0\}$ such that*

$$d\omega^i = \omega^j \wedge \omega_j{}^i, \tag{2.3}$$

$$dg_{ij} = g_{kj}\omega_i{}^k + g_{ik}\omega_j{}^k + 2C_{ijk}\omega^{n+k}, \tag{2.4}$$

where

$$\omega^{n+i} := dy^i + y^j\omega_j{}^i, \tag{2.5}$$

where $\mathcal{Y} =: y^i\mathbf{e}_i$, $g_{ij} := \mathcal{G}(\mathbf{e}_i, \mathbf{e}_j)$ *and* $C_{ijk} := \mathcal{C}(\mathbf{e}_i, \mathbf{e}_j, \mathbf{e}_k)$.

Proof: We will prove the theorem in a standard local coordinate system (x^i, y^i) in TM_o. Taking $\mathbf{e}_i = \partial_i$ and $\omega^i := dx^i$. The local 1-forms $\omega_j{}^i$ can be expressed as

$$\omega_j{}^i := \Gamma^i_{jk} dx^k + \Pi^i_{jk} dy^k.$$

(2.3) is equivalent to the following equation

$$0 = d^2 x^i = dx^j \wedge \left(\Gamma^i_{jk} dx^k + \Pi^i_{jk} dy^k \right).$$

This implies that $\Pi^i_{jk} = 0$ and $\Gamma^i_{jk} = \Gamma^i_{kj}$. Thus $\omega_j{}^i = \Gamma^i_{jk} dx^k$. Let

$$N^i_j := y^m \Gamma^i_{mj}. \tag{2.6}$$

(2.4) becomes

$$dg_{ij} = g_{im} \Gamma^m_{jl} dx^l + g_{mj} \Gamma^m_{il} dx^l + 2C_{ijm} \left(dy^m + N^m_l dx^l \right). \tag{2.7}$$

Since

$$C_{ijk} = \frac{1}{4} [F^2]_{y^i y^j y^k} = \frac{1}{2} \frac{\partial g_{ij}}{\partial y^k},$$

(2.7) is further reduced to the following

$$\frac{\partial g_{ij}}{\partial x^l} = g_{im} \Gamma^m_{jl} + g_{mj} \Gamma^m_{il} + 2C_{ijm} N^m_l. \tag{2.8}$$

Permutating the indices, one obtains from (2.8) that

$$\frac{\partial g_{jl}}{\partial x^i} = g_{jm} \Gamma^m_{il} + g_{ml} \Gamma^m_{ij} + 2C_{jml} N^m_i, \tag{2.9}$$

$$\frac{\partial g_{li}}{\partial x^j} = g_{ml} \Gamma^m_{ij} + g_{im} \Gamma^m_{jl} + 2C_{iml} N^m_j. \tag{2.10}$$

Adding (2.9) and (2.10), then subtracting (2.8) yield

$$\frac{\partial g_{jl}}{\partial x^i} + \frac{\partial g_{li}}{\partial x^j} - \frac{\partial g_{ij}}{\partial x^l} = 2g_{ml} \Gamma^m_{ij} + 2C_{jml} N^m_i + 2C_{iml} N^m_j - 2C_{ijm} N^m_l.$$

From the above identity, one obtains

$$\Gamma^k_{ij} = \frac{1}{2} g^{kl} \left\{ \frac{\partial g_{jl}}{\partial x^i} + \frac{\partial g_{li}}{\partial x^j} - \frac{\partial g_{ij}}{\partial x^l} \right\}$$
$$- g^{kl} \left\{ C_{jml} N^m_i + C_{iml} N^m_j - C_{ijm} N^m_l \right\}. \tag{2.11}$$

To determine Γ^i_{jk} completely, one needs to express N^i_j in terms of F. Contracting (2.11) with y^i yields

$$N^k_j = \frac{1}{2}g^{kl}\left\{\frac{\partial g_{jl}}{\partial x^i} + \frac{\partial g_{li}}{\partial x^j} - \frac{\partial g_{ij}}{\partial x^l}\right\}y^i - 2g^{kl}C_{jml}G^m, \qquad (2.12)$$

where

$$G^i := \frac{1}{2}N^i_j y^j = \frac{1}{2}\Gamma^i_{jk}y^j y^k.$$

Contracting (2.12) with $\frac{1}{2}y^j$ yields

$$G^i = \frac{1}{4}g^{il}\left\{\frac{\partial g_{jl}}{\partial x^k} + \frac{\partial g_{lk}}{\partial x^j} - \frac{\partial g_{jk}}{\partial x^l}\right\}y^j y^k. \qquad (2.13)$$

Plugging (2.13) into (2.12), one obtains a formula for N^i_j expressed in terms of g_{ij}. Plugging them into (2.11) yields a formula for Γ^i_{jk} expressed in terms of g_{ij}.   Q.E.D.

With the Chern connection forms $\{\omega_j{}^i\}$ with respect to a local frame $\{\mathbf{e}_i\}$ for π^*TM, one can define a linear connection ∇ on π^*TM by

$$\nabla X := \left\{dX^i + X^j\omega_j{}^i\right\} \otimes \mathbf{e}_i,$$

where $X = X^i(x, y)\mathbf{e}_i \in C^\infty(\pi^*TM)$. Clearly, ∇ is a well-defined linear connection on π^*TM. It is also torsion-free in the following sense:

$$\nabla_{\hat{X}}\rho(\hat{Y}) - \nabla_{\hat{Y}}\rho(\hat{X}) = \rho[\hat{X}, \hat{Y}], \qquad \hat{X}, \hat{Y} \in C^\infty(T(TM_o)),$$

where $\rho : T(TM_o) \to \pi^*TM$ is a vector bundle map defined by

$$\rho\left(\frac{\partial}{\partial x^i}|_{(x,y)}\right) = \partial_i|_x, \qquad \rho\left(\frac{\partial}{\partial y^i}|_{(x,y)}\right) = 0.$$

∇ is called the *Chern connection*.

Using

$$[F^2]_{x^l} = \frac{\partial g_{ij}}{\partial x^l}y^i y^j, \qquad [F^2]_{x^k y^l}y^k = 2\frac{\partial g_{il}}{\partial x^k}y^i y^k,$$

we can rewrite G^i in (2.13) as follows,

$$G^i = \frac{1}{4}g^{il}\left\{[F^2]_{x^k y^l}y^k - [F^2]_{x^l}\right\}. \qquad (2.14)$$

Using

$$[F^2]_{x^l} = 2FF_{x^l}, \quad [F^2]_{x^k y^l} y^k = \frac{2F_{x^k} y^k}{F} g_{ml} y^m + 2FF_{x^k y^l} y^k,$$

we obtain

$$G^i = Py^i + Q^i, \tag{2.15}$$

where

$$P := \frac{F_{x^k} y^k}{2F}, \quad Q^i := \frac{F}{2} g^{il} \left\{ F_{x^k y^l} y^k - F_{x^l} \right\}.$$

Let

$$G := y^i \frac{\partial}{\partial x^i} - 2G^i \frac{\partial}{\partial y^i}, \tag{2.16}$$

where $G^i = G^i(x, y)$ are defined in (2.13) with the following homogeneity

$$G^i(x, \lambda y) = \lambda^2 G^i(x, y), \qquad \lambda > 0. \tag{2.17}$$

G is a well-defined vector field on TM_o. We call G the *spray* induced by F and G^i the *spray coefficients* of F.

Any C^∞ vector field G on TM_o in the form (2.16) with homogeneity (2.17) is called a *spray* on M. Not all spray can be induced by a Finsler metric. Thus sprays are more general geometric structures. See [86] for discussions on sprays.

The local functions N_j^i defined in (2.6) or (2.12) are called the *connection coefficients*. Differentiating (2.13) with respect to y^j yields

$$N_j^i = \frac{\partial G^i}{\partial y^j}. \tag{2.18}$$

Thus N_j^i depend on G^i only. Let

$$\frac{\delta}{\delta x^i} := \frac{\partial}{\partial x^i} - N_i^j \frac{\partial}{\partial y^j}. \tag{2.19}$$

Then $HTM := \text{span}\{\frac{\delta}{\delta x^i}\}$ is a well-defined subbundle of $T(TM_o)$ and $T(TM_o) = HTM \oplus VTM$. Note that $2G^i = N_j^i y^j$. Thus $G = y^i \frac{\delta}{\delta x^i}$ is a special section of HTM.

The local functions Γ^i_{jk} in (2.11) are called the *Christoffel symbols* of the Chern connection. N^i_j and Γ^i_{jk} are related by (2.6). However, differentiating N^i_j with respect to y^k does not give Γ^i_{jk}. Let

$$L^i{}_{jk} := \frac{\partial^2 G^i}{\partial y^j \partial y^k} - \Gamma^i_{jk}. \qquad (2.20)$$

We obtain a tensor $\mathcal{L} = L^i{}_{jk}\partial_i \otimes dx^j \otimes dx^k$. We call $\mathcal{L}$ the *Landsberg tensor*. Note that $L^i{}_{jk}$ are symmetric in j, k. The Landsberg tensor is an important quantity in Finsler geometry. Let

$$J^i := g^{jk}L^i{}_{jk}. \qquad (2.21)$$

Then $\mathcal{J} := J^i\partial_i$ is called the *mean Landsberg tensor*.

It is easy to verify that

$$L^i{}_{jk} = y^m \frac{\partial \Gamma^i_{mk}}{\partial y^j}. \qquad (2.22)$$

Thus if $\Gamma^i_{jk} = \Gamma^i_{jk}(x)$ are functions of x only, then $L^i{}_{jk} = 0$.

Now let us take a look at the Christoffel symbols of Riemannian metrics. Let $F = \sqrt{g_{ij}(x)y^i y^j}$ be a Riemannian metric on a manifold M. By (2.13), the spray coefficients $G^i = G^i(x, y)$ of F are given by

$$G^i = \frac{1}{4}g^{il}(x)\left\{\frac{\partial g_{jl}}{\partial x^k}(x) + \frac{\partial g_{kl}}{\partial x^j}(x) - \frac{\partial g_{jk}}{\partial x^l}(x)\right\}y^j y^k, \qquad (2.23)$$

where $(g^{ij}(x)) := (g_{ij}(x))^{-1}$. By (2.6) and (2.11), the connection coefficients N^i_j and the Christoffel symbols Γ^i_{jk} are given by

$$N^i_j = \frac{1}{2}g^{il}(x)\left\{\frac{\partial g_{jl}}{\partial x^k}(x) + \frac{\partial g_{kl}}{\partial x^j}(x) - \frac{\partial g_{jk}}{\partial x^l}(x)\right\}y^k,$$

$$\Gamma^i_{jk} = \frac{1}{2}g^{il}(x)\left\{\frac{\partial g_{jl}}{\partial x^k}(x) + \frac{\partial g_{kl}}{\partial x^j}(x) - \frac{\partial g_{jk}}{\partial x^l}(x)\right\}.$$

Note that the Christoffel symbols $\Gamma^i_{jk} = \Gamma^i_{jk}(x)$ are functions of x only. Thus $L^i{}_{jk} = 0$.

Definition 2.1.2   A Finsler metric F on a manifold M is called a *Berwald metric* if in any standard local coordinate system (x^i, y^i) in TM_o, the Christoffel symbols $\Gamma^i_{jk} = \Gamma^i_{jk}(x)$ are functions of $x \in M$ only, in which case, $G^i = \frac{1}{2}\Gamma^i_{jk}(x)y^j y^k$ are quadratic in $y = y^i \frac{\partial}{\partial x^i}|_x$. F is called a *Landsberg metric* if $L^i{}_{jk} = 0$.

As shown above, Riemannian are Berwald metrics. There are many non-Riemannian Berwald metrics. See Section 4.3 below. G. Landsberg first studied Landsberg metrics [61], [62]. Since then, Landsberg metrics frequently appear in literatures. By (2.20) or (2.22), we obtain the following

Proposition 2.1.3 *Every Berwald metric is a Landsberg metric.*

It is still an open problem whether or not every Landsberg metric is a Berwald metric. So far no counter-example has been found yet.

Let F be a Berwald metric on a manifold M. By definition, in any standard local coordinate system (x^i, y^i) in TM_o, the Christoffel symbols $\Gamma^i_{jk} = \Gamma^i_{jk}(x)$ are local functions of $x \in M$ only. Let $\theta_j{}^i := \Gamma^i_{jk}(x)dx^k$. Define

$$DX := \left\{ dX^i + X^j \theta^i_j \right\} \otimes \frac{\partial}{\partial x^i}, \qquad (2.24)$$

where $X = X^j \frac{\partial}{\partial x^j} \in C^\infty(TM)$. Clearly, D is a linear connection on TM. It is also torsion-free in the following sense:

$$D_X Y - D_Y X = [X, Y], \qquad X, Y \in C^\infty(TM).$$

D is called the *Levi-Civita connection* of F.

Consider a Riemannian metric $g = g_{ij}(x)dx^i \otimes dx^j$ on M. Let $\Gamma^i_{jk} = \Gamma^i_{jk}(x)$ denote the Christoffel symbols. Let D be the Levi-Civita connection defined by (2.24) using $\theta_j{}^i := \Gamma^i_{jk}(x)dx^k$. One can easily verify that

$$d\theta^i = \theta^j \wedge \theta_j{}^i \qquad (2.25)$$

$$dg_{ij} = g_{ik}\theta_j{}^k + g_{jk}\theta_i{}^k. \qquad (2.26)$$

(2.25) and (2.26) are equivalent to the following equations in the index-free form:

$$D_X Y - D_Y X = [X, Y], \qquad (2.27)$$

$$y[g(X, Y)] = g(D_y X, Y) + g(X, D_y Y), \qquad (2.28)$$

where $y \in T_x M$ and $X, Y \in C^\infty(TM)$.

One can show that for a Riemannian metric g on a manifold M. D is the unique linear connection on TM satisfying (2.27) and (2.28).

2.2   Structure Equations

In this section we are going to use the Chern connection to introduce several notions of curvatures for Finsler metrics.

Let (M, F) be an n-dimensional Finsler manifold. Let $\{\mathbf{e}_i\}$ be an arbitrary local frame for π^*TM and $\{\omega^i\}$ the dual coframe for $\pi^*T^*M \cong HTM$. Express $\mathcal{Y}, \mathcal{G}$ and $\mathcal{C}$ by

$$\mathcal{Y} = y^i \mathbf{e}_i, \qquad \mathcal{G} = g_{ij}\omega^i \otimes \omega^j, \qquad \mathcal{C} = C_{ijk}\omega^i \otimes \omega^j \otimes \omega^k.$$

See (2.1) and (2.2) for definitions.

According to Theorem 2.1.1, the Chern connection forms $\{\omega_j{}^i\}$ with respect to $\{\mathbf{e}_i\}$ are uniquely determined by

$$d\omega^i = \omega^j \wedge \omega_j{}^i, \tag{2.29}$$

$$dg_{ij} = g_{ik}\omega_j{}^k + g_{kj}\omega_i{}^k + 2C_{ijk}\omega^{n+k}, \tag{2.30}$$

where

$$\omega^{n+i} := dy^i + y^j \omega_j{}^i. \tag{2.31}$$

Clearly, $\{\omega^i, \omega^{n+i}\}$ is a local coframe for $T^*(TM_o)$. The curvature forms $\Omega_j{}^i$ are defined by

$$\Omega_j{}^i := d\omega_j{}^i - \omega_j{}^k \wedge \omega_k{}^i. \tag{2.32}$$

Differentiating (2.29), one obtains

$$\begin{aligned}
d^2\omega^i &= d\omega^j \wedge \omega_j{}^i - \omega^j \wedge d\omega_j{}^i \\
&= \left\{ \omega^m \wedge \omega_m{}^j \right\} \wedge \omega_j{}^i - \omega^j \wedge \left\{ \Omega_j{}^i + \omega_j{}^m \wedge \omega_m{}^i \right\} \\
&= -\omega^j \wedge \Omega_j{}^i.
\end{aligned}$$

Since $d^2\omega^i = 0$, one obtains the following identity:

$$\omega^j \wedge \Omega_j{}^i = 0.$$

The above identity is called the *first Bianchi identity*. $\Omega_j{}^i$ can be expressed in terms of $\omega^i \wedge \omega^j$, $\omega^i \wedge \omega^{n+j}$ and $\omega^{n+i} \wedge \omega^{n+j}$. By the first Bianchi identity, we can express $\Omega_j{}^i$ by

$$\Omega_j{}^i = \frac{1}{2}R_j{}^i{}_{kl}\omega^k \wedge \omega^l + P_j{}^i{}_{kl}\omega^k \wedge \omega^{n+l}, \tag{2.33}$$

where

$$R_j{}^i{}_{kl} + R_j{}^i{}_{lk} = 0, \tag{2.34}$$

$$R_j{}^i{}_{kl} + R_k{}^i{}_{lj} + R_l{}^i{}_{jk} = 0, \tag{2.35}$$

$$P_j{}^i{}_{kl} = P_k{}^i{}_{jl}. \tag{2.36}$$

Let

$$\Omega^i := d\omega^{n+i} - \omega^{n+j} \wedge \omega_j{}^i. \tag{2.37}$$

Differentiating (2.31), we obtain

$$\begin{aligned}
\Omega^i &= d\omega^{n+i} - \omega^{n+j} \wedge \omega_j{}^i \\
&= d^2 y^i + dy^j \wedge \omega_j{}^i + y^j d\omega_j{}^i - \omega^{n+j} \wedge \omega_j{}^i \\
&= \left\{ \omega^{n+j} - y^m \omega_m{}^j \right\} \wedge \omega_j{}^i + y^j \left\{ \Omega_j{}^i + \omega_j{}^m \wedge \omega_m{}^i \right\} - \omega^{n+j} \wedge \omega_j{}^i \\
&= y^j \Omega_j{}^i.
\end{aligned}$$

That is,

$$\Omega^i = y^j \Omega_j{}^i. \tag{2.38}$$

Ω^i is the essential part of $\Omega_j{}^i$. By (2.38), one can express Ω^i in the following form

$$\Omega^i = \frac{1}{2} R^i{}_{kl} \omega^k \wedge \omega^l - L^i{}_{kl} \omega^k \wedge \omega^{n+l}, \tag{2.39}$$

where

$$R^i{}_{kl} + R^i{}_{lk} = 0,$$

and

$$R^i{}_{kl} := y^j R_j{}^i{}_{kl}, \qquad L^i{}_{kl} := -y^j P_j{}^i{}_{kl}. \tag{2.40}$$

Clearly,

$$y^k R^i{}_{kl} = 0, \qquad y^k L^i{}_{kl} = 0. \tag{2.41}$$

Setting

$$R^i{}_k := R^i{}_{kl} y^l = y^j R_j{}^i{}_{kl} y^l, \tag{2.42}$$

we obtain the *Riemann tensor*.

$$\mathcal{R} := R^i{}_k \mathbf{e}_i \otimes \omega^k. \tag{2.43}$$

The Riemann tensor $\mathcal{R}$ has the following properties.

$$R^i{}_k y^k = 0, \qquad R_{ij} = R_{ji}, \tag{2.44}$$

where $R_{ij} := g_{im} R^m_j$. The first identity follows from (2.41) and the second identity will be proved using Bianchi identities in Section 2.4 below.

In a standard local coordinate system (x^i, y^i) in TM_o, the natural local coframe $\{\omega^i, \omega^{n+i}\}$ on TM_o are given by

$$\omega^i = dx^i, \qquad \omega^{n+i} = \delta y^i := dy^i + N^i_j dx^j,$$

where $N^i_j = y^m \Gamma^i_{mj}$ are defined in (2.6). Plugging $\omega_j{}^i = \Gamma^i_{jk} dx^k$ into (2.32) yields

$$R_j{}^i{}_{kl} = \frac{\delta \Gamma^i_{jl}}{\delta x^k} - \frac{\delta \Gamma^i_{jk}}{\delta x^l} + \Gamma^i_{ks} \Gamma^s_{jl} - \Gamma^s_{jk} \Gamma^i_{ls}, \tag{2.45}$$

$$P_j{}^i{}_{kl} = -\frac{\partial \Gamma^i_{jk}}{\partial y^l}, \tag{2.46}$$

where $\frac{\delta}{\delta x^i}$ are defined in (2.19).

It follows from (2.40) and (2.45) that

$$R^i{}_{kl} = \frac{\partial N^i_l}{\partial x^k} - \frac{\partial N^i_k}{\partial x^l} + N^s_l \frac{\partial N^i_k}{\partial y^s} - N^s_k \frac{\partial N^i_l}{\partial y^s}. \tag{2.47}$$

Observe that

$$y^j \frac{\partial \Gamma^i_{jk}}{\partial y^l} = \frac{\partial (\Gamma^i_{jk} y^j)}{\partial y^l} - \Gamma^i_{kl} = \frac{\partial N^i_k}{\partial y^l} - \Gamma^i_{kl} = \frac{\partial^2 G^i}{\partial y^k \partial y^l} - \Gamma^i_{kl}.$$

We obtain from (2.40) and (2.46) that

$$L^i{}_{kl} = y^j \frac{\partial \Gamma^i_{jk}}{\partial y^l} = \frac{\partial^2 G^i}{\partial y^k \partial y^l} - \Gamma^i_{kl}. \tag{2.48}$$

By (2.48), one can see that $L^i{}_{jk}$ defined in (2.40) are just those defined in (2.20).

Since $N^i_j = \frac{\partial G^i}{\partial y^j}$, $R^i{}_{kl}$ can be expressed directly in terms of G^i. Contracting (2.47) with y^l and using (2.18) we obtain

$$R^i{}_k = 2\frac{\partial G^i}{\partial x^k} - y^j \frac{\partial^2 G^i}{\partial x^j \partial y^k} + 2G^j \frac{\partial^2 G^i}{\partial y^j \partial y^k} - \frac{\partial G^i}{\partial y^j}\frac{\partial G^j}{\partial y^k}. \tag{2.49}$$

By (2.47) and (2.49), one can verify the following identity directly.

$$R^i{}_{kl} = \frac{1}{3}\left\{ \frac{\partial R^i{}_k}{\partial y^l} - \frac{\partial R^i{}_l}{\partial y^k} \right\}. \tag{2.50}$$

Thus, $R^i{}_k$ and $R^i{}_{kl}$ can determine each other directly by vertical differentiation. According to (2.49), the Riemann tensor $\mathcal{R}$ depends only on the spray $G = y^i \frac{\partial}{\partial x^i} - 2G^i \frac{\partial}{\partial y^i}$ of the Finsler metric.

2.3 Finsler Metrics of Constant Flag Curvature

The Riemann tensor $\mathcal{R} := R^i{}_k \mathbf{e}_i \otimes \omega^k$ is the most important geometric quantity in Finsler geometry. It measures the curvature of a Finsler manifold. We are particularly interested in evenly curved Finsler manifolds.

Let

$$h^i{}_k := \delta^i_k - F^{-2}g_{kq}y^q y^i.$$

We obtain the *angular tensor* $h := h^i{}_k \mathbf{e}_i \otimes \omega^k$. Note that $h^i{}_k$ has the following properties similar to that of $R^i{}_k$, i.e.,

$$h^i{}_k y^k = 0.$$

It is natural to compare $\mathcal{R}$ with h. A Finsler metric F has *constant flag curvature* μ if $\mathcal{R} = \mu F^2 h$, i.e.,

$$R^i{}_k = \mu F^2 h^i{}_k = \mu\left\{ F^2\delta^i_k - g_{kq}y^q y^i \right\}. \tag{2.51}$$

By (2.42) and (2.50), we can see that (2.51) is equivalent to the following equation:

$$R^i{}_{kl} = \mu\left\{ g_{lp}y^p \delta^i_k - g_{kq}y^q \delta^i_l \right\}. \tag{2.52}$$

It follows from (2.52) that

$$\frac{\partial}{\partial y^j}(R^i{}_{kl}) = \mu\left\{ g_{jl}\delta^i_k - g_{jk}\delta^i_l \right\}.$$

Assume that F is a Berwald metric. By definition, in any standard local coordinate system (x^i, y^i), $\Gamma^i_{jk} = \Gamma^i_{jk}(x)$ are functions of x only. By (2.45),

$$R_j{}^i{}_{kl} = \frac{\partial \Gamma^i_{jl}}{\partial x^k} - \frac{\partial \Gamma^i_{jk}}{\partial x^l} + \Gamma^i_{ks}\Gamma^s_{jl} - \Gamma^s_{jk}\Gamma^i_{ls}. \qquad (2.53)$$

We see that $R_j{}^i{}_{kl} = R_j{}^i{}_{kl}(x)$ are local functions of x only. In this case, we have

$$R^i{}_{kl} = R_j{}^i{}_{kl}y^j, \qquad R_j{}^i{}_{kl} = \frac{\partial}{\partial y^j}(R^i{}_{kl}).$$

Thus for a Berwald metric F, it is of constant flag curvature μ if and only if

$$R_j{}^i{}_{kl} = \mu\Big\{g_{jl}\delta^i_k - g_{jk}\delta^i_l\Big\}.$$

Note that

$$R_j{}^m{}_{ml} = (n-1)\mu g_{jl}.$$

Thus if $\mu \neq 0$, then g_{jl} are independent of $y \in T_x M$. We conclude that any Berwald metric of non-zero constant flag curvature must be Riemannian. The case when $\mu = 0$ will be discussed in Theorem 2.3.2 below.

Below are some examples of Finsler metrics of constant flag curvature.

Example 2.3.1   For a constant μ, let $F = \alpha_\mu$ be the Riemannian metric defined in (1.12),

$$F := \frac{\sqrt{|y|^2 + \mu(|x|^2|y|^2 - \langle x, y\rangle^2)}}{1 + \mu|x|^2}, \qquad y \in T_x B^n(r_\mu) \cong R^n, \qquad (2.54)$$

where $r_\mu := 1/\sqrt{-\mu}$ if $\mu < 0$ and $r_\mu := +\infty$ if $\mu \geq 0$. First we have

$$F_{x^k}y^k = -\frac{2\mu\langle x, y\rangle}{1 + \mu|x|^2}F,$$

and

$$F_{x^k y^l}y^k - F_{x^l} = [F_{x^k}y^k]_{y^l} - 2F_{x^l} = 0.$$

By (2.15), one obtains

$$G^i = -\frac{\mu\langle x, y\rangle}{1 + \mu|x|^2}\, y^i. \qquad (2.55)$$

Plugging them into (2.49) yields that

$$R^i_{\ k} = \mu F^2 h^i_{\ k}.$$

Thus F has constant flag curvature μ.

One of the fundamental theorems in Riemannian geometry states that *every Riemannian metric of constant flag curvature μ can be locally expressed as (2.54) in some local coordinate system.* This is a special case of E. Cartan's solution to the local equivalence problem.

A Finsler metric F is said to be *flat* if the induced spray G is flat. More precisely, at any point x in M, there is a standard local coordinate system (x^i, y^i) in TM in which the spray coefficients $G^i = 0$. F is said to be *locally Minkowskian* if at any point $x \in M$, there is a standard local coordinate system (x^i, y^i) in TM in which $F = F(y)$ is a function of $(y^i) \in \mathbb{R}^n$ only. Clearly, any locally Minkowskian metric is flat. The converse is true too. Suppose that a Finsler metric is flat, i.e., $G^i = 0$ in a local coordinate system (x^i, y^i). It follows from (2.8) that $F^2 = g_{ij} y^i y^j$ satisfies

$$[F^2]_{x^k} = \frac{\partial g_{ij}}{\partial x^k} y^i y^j = 2 y^i g_{im} N^m_k = 0.$$

Thus F is locally Minkowskian.

For a Berwald metric, the flatness can be characterized by the curvature equation $\mathcal{R} = 0$.

Theorem 2.3.2 (Berwald) *Let F be a Berwald metric on a manifold. F is flat if and only if $\mathcal{R} = 0$.*

Here is a sketch of the proof. Assume that F is a Berwald metric. In this case, the Levi-Civita connection D is defined in (2.24). D is a torsion-free linear connection on TM. The curvature condition $\mathcal{R} = 0$ implies that D is flat. Then there is a local coordinate system (x^i) in which, $\Gamma^i_{jk} = 0$ (see [36]), hence F is flat.   Q.E.D.

There are many non-Berwaldian Finsler metrics with $\mathcal{R} = 0$. See the following example.

Example 2.3.3   Let $B^n(1) \subset \mathbb{R}^n$ be the standard unit ball and define

$$F := \frac{\left(\sqrt{|y|^2 - (|x|^2|y|^2 - \langle x, y \rangle^2)} + \langle x, y \rangle\right)^2}{(1 - |x|^2)^2 \sqrt{|y|^2 - (|x|^2|y|^2 - \langle x, y \rangle^2)}}, \qquad (2.56)$$

where $y \in T_x B^n(1) \cong \mathbb{R}^n$. It is easy to verify that $F = F(x, y)$ satisfies the following equations

$$F_{x^k y^l} y^k = F_{x^l}$$

and

$$F_{x^k} y^k = 2\Theta F,$$

where

$$\Theta = \frac{\sqrt{|y|^2 - (|x|^2|y|^2 - \langle x, y \rangle^2)} + \langle x, y \rangle}{1 - |x|^2}$$

is the Funk metric on $B^n(1)$ which is defined in (1.34). Then by (2.15), we obtain that $G^i = \Theta y^i$. It is easy to verify that $\Theta = \Theta(x, y)$ satisfies (1.38),

$$\Theta_{x^k} = \Theta \Theta_{y^k}.$$

We have

$$\Theta_{x^j y^k} y^j = \Theta_{x^k}, \qquad y^j \Theta_{x^j} = \Theta^2, \qquad y^j \Theta_{y^j y^k} = 0.$$

Then by (2.49) and the above identities, we obtain

$$R^i{}_k = 2\Theta_{x^k} y^i - y^j \left\{ \Theta_{x^j y^k} y^i + \Theta_{x^j} \delta^i_k \right\}$$

$$+ 2\Theta y^j \left\{ \Theta_{y^j y^k} y^i + \Theta_{y^k} \delta^i_j + \Theta_{y^j} \delta^i_k \right\}$$

$$- \left\{ \Theta_{y^j} y^i + \Theta \delta^i_j \right\} \left\{ \Theta_{y^k} y^j + \Theta \delta^j_k \right\}$$

$$= \left\{ 2\Theta_{x^k} - \Theta_{x^j y^k} y^j + 2\Theta \Theta_{y^j y^k} y^j - \Theta \Theta_{y^k} \right\} y^i$$

$$+ \left\{ \Theta^2 - y^j \Theta_{x^j} \right\} \delta^i_k = 0.$$

Thus F has zero flag curvature. The Finsler metric in (2.56) is constructed by L. Berwald [17]. See Example 3.4.7 for a different proof. See also [89] and [75] for related discussion.

2.4 Bianchi Identities

In this section we will employ the exterior differentiation method to derive the relationship among curvatures.

Let (M, F) be an n-dimensional Finsler manifold. Let $\{e_i\}$ be a local frame for $\pi^* TM$, $\{\omega^i, \omega^{n+i}\}$ be the corresponding local coframe for $T^*(TM_o)$ and $\{\omega_j{}^i\}$ be the set of local Chern connection forms with respect to $\{e_i\}$. For a scalar function f on $TM \setminus \{0\}$, define $f_{|m}$ and $f_{\cdot m}$ by

$$df = f_{|k}\omega^k + f_{\cdot k}\omega^{n+k}. \tag{2.57}$$

There is a canonical way to define the covariant derivatives of a tensor on TM_o using the Chern connection. For example, if $T = T_{ij}\omega^i \otimes \omega^j$, $T_{ij|k}$ and $T_{ij\cdot k}$ are defined by

$$T_{ij|k}\omega^k + T_{ij\cdot k}\omega^{n+k} := dT_{ij} - T_{kj}\omega_i{}^k - T_{ik}\omega_j{}^k.$$

In a standard local coordinate system (x^i, y^i), the coefficients, $T_{ij} = T_{ij}(x, y)$, are local functions of (x^i, y^i), where $y = y^i \frac{\partial}{\partial x^i}|_x$. $T_{ij|k}$ and $T_{ij\cdot k}$ are given by

$$T_{ij|k} = \frac{\partial T_{ij}}{\partial x^k} - T_{sj}\Gamma^s_{ik} - T_{is}\Gamma^s_{jk} - T_{ij\cdot s}N^s_k,$$
$$T_{ij\cdot k} = \frac{\partial T_{ij}}{\partial y^k}.$$

The covariant differentiation satisfies the product rule. For example, for $S = S_{ij}\omega^i \otimes \omega^j$ and $T = T_k\omega^k$,

$$(S_{ij}T_k)_{|l} = S_{ij|l}T_k + S_{ij}T_{k|l}.$$

For the fundamental tensor $\mathcal{G} = g_{ij}\omega^i \otimes \omega^j$, it follows from (2.30) that

$$g_{ij|k} = 0, \qquad g_{ij\cdot k} = 2C_{ijk}.$$

For the canonical section $\mathcal{Y} = y^i e_i$, it follows from (2.31) that

$$y^i_{|k} = 0, \qquad y^i_{\cdot k} = \delta^i_k.$$

Thus

$$[F^2]_{|m} = 2FF_{|m} = g_{ij|m}y^i y^j = 0. \tag{2.58}$$

Recall the curvature forms in (2.39)

$$\Omega^i := d\omega^{n+i} - \omega^{n+j} \wedge \omega_j{}^i = \frac{1}{2}R^i{}_{kl}\omega^k \wedge \omega^l - L^i{}_{kl}\omega^k \wedge \omega^{n+l}. \qquad (2.59)$$

Differentiating (2.59) yields the following *second Bianchi identities*,

$$d\Omega^i = -\Omega^j \wedge \omega_j{}^i + \omega^{n+j} \wedge \Omega_j{}^i. \qquad (2.60)$$

It follows from (2.60) that

$$R^i{}_{kl|j} + R^i{}_{lj|k} + R^i{}_{jk|l} = -L^i{}_{jm}R^m{}_{kl} - L^i{}_{km}R^m{}_{lj} - L^i{}_{lm}R^m{}_{jk}, \cdot \quad (2.61)$$

$$R_j{}^i{}_{kl} = R^i{}_{kl \cdot j} + L^i{}_{kj|l} - L^i{}_{lj|k} + L^i{}_{lm}L^m{}_{kj} - L^i{}_{km}L^m{}_{lj}. \qquad (2.62)$$

Contracting (2.61) with y^l yields

$$R^i{}_{k|j} - R^i{}_{j|k} + R^i{}_{jk|m}y^m = L^i{}_{km}R^m{}_j - L^i{}_{jm}R^m{}_k. \qquad (2.63)$$

We now use (2.30) to find other relationship between the curvature tensors and the Finsler metric. Differentiating (2.30) yields

$$0 = g_{ik}\Omega_j{}^k + g_{kj}\Omega_i{}^k + 2(C_{ijk|l}\omega^l + C_{ijk \cdot l}\omega^{n+l}) \wedge \omega^{n+k} + 2C_{ijk}\Omega^k.$$

This gives the following equations:

$$R_{jikl} + R_{ijkl} + 2C_{ijm}R^m{}_{kl} = 0, \qquad (2.64)$$

$$P_{jikl} + P_{ijkl} + 2C_{ijl|k} - 2C_{ijm}L^m{}_{kl} = 0, \qquad (2.65)$$

where $R_{jikl} := g_{im}R_j{}^m{}_{kl}$ and $P_{jikl} := g_{im}P_j{}^m{}_{kl}$.

From (2.34) and (2.35), one obtains

$$2(R_{klji} - R_{jikl}) = (R_{klji} + R_{lkji}) - (R_{jikl} + R_{ijkl})$$
$$+(R_{kilj} + R_{iklj}) + (R_{ljki} + R_{jlki})$$
$$+(R_{iljk} + R_{lijk}) + (R_{jkil} + R_{kjil}).$$

Applying (2.64) to the above identity, one obtains

$$R_{klji} - R_{jikl} = C_{klm}R^m{}_{ji} - C_{jim}R^m{}_{kl} + C_{kim}R^m{}_{lj}$$
$$+C_{ljm}R^m{}_{ki} + C_{ilm}R^m{}_{jk} + C_{jkm}R^m{}_{il}. \qquad (2.66)$$

Let $R_{ij} := g_{ik}R^k{}_j$. Contracting (2.66) with y^j and y^k yields that

$$R_{ij} = R_{ji}.$$

This verifies the identity (2.44).

We now discuss the Landsberg tensor $\mathcal{L} = L^i{}_{kl}\mathbf{e}_i \otimes \omega^k \otimes \omega^l$. Let

$$L_{ijk} := g_{im}L^m_{jk} = -y^m P_{mijk}.$$

By (2.36) and (2.65), one obtains

$$
\begin{aligned}
L_{ijk} &= -\frac{1}{2}y^m P_{mijk} - \frac{1}{2}y^m P_{jimk} \\
&= \frac{1}{2}y^m P_{imjk} + \frac{1}{2}y^m P_{jimk} + C_{ijk|m}y^m \\
&= \frac{1}{2}y^m P_{imjk} + \frac{1}{2}y^m P_{mijk} + C_{ijk|m} \\
&= C_{ijk|m}y^m.
\end{aligned}
\tag{2.67}
$$

Thus L_{ijk} is symmetric in i, j, k. Let $J_i := g_{im}J^m = g^{jk}L_{ijk}$. Contracting (2.67) with g^{jk} yields

$$J_i = I_{i|m}y^m. \tag{2.68}$$

By virtue of (2.67) and (2.68),

$$L_{ijk|m}y^m = C_{ijk|p|q}y^p y^q, \qquad J_{k|m}y^m = I_{k|p|q}y^p y^q.$$

Lemma 2.4.1 ([72], [74])

$$
\begin{aligned}
C_{ijk|p|q}y^p y^q + C_{ijm}R^m_k &= -\frac{1}{3}g_{im}R^m_{k\cdot j} - \frac{1}{3}g_{jm}R^m_{k\cdot i} \\
&\quad -\frac{1}{6}g_{im}R^m_{j\cdot k} - \frac{1}{6}g_{jm}R^m_{i\cdot k}
\end{aligned}
\tag{2.69}
$$

and

$$I_{k|p|q}y^p y^q + I_m R^m_k = -\frac{1}{3}\Big\{ 2R^m_{k\cdot m} + R^m_{m\cdot k} \Big\}. \tag{2.70}$$

Proof. It follows from (2.62) that

$$R_j{}^i{}_{km}y^m = R^i{}_{km\cdot j}y^m + L_{ijk|m}y^m = R^i{}_{k\cdot j} + R^i{}_{jk} + L^i{}_{jk|m}y^m.$$

Thus

$$R_{jikm}y^m = g_{im}\Big\{ R^m_{k\cdot j} + R^m_{jk} \Big\} + L_{ijk|m}y^m.$$

Structure Equations

By the above formulas for $R_{jikm}y^m$ and (2.64), one obtains

$$L_{ijk|m}y^m + C_{ijm}R^m_k = -\frac{1}{2}g_{im}\left\{R^m_{k\cdot j} + R^m_{jk}\right\}$$

$$-\frac{1}{2}g_{jm}\left\{R^m_{k\cdot i} + R^m_{ik}\right\}. \qquad (2.71)$$

Contracting (2.71) with g^{ij}

$$J_{k|m}y^m + I_m R^m_k = -\left\{R^m_{k\cdot m} + R^m_{m\cdot k}\right\}. \qquad (2.72)$$

One can rewrite (2.50) as follows

$$R^i_{kl} = \frac{1}{3}\left\{R^i_{k\cdot l} - R^i_{l\cdot k}\right\}. \qquad (2.73)$$

(Note: the reader can derive other Bianchi identities and verify (2.73) using these identities, then express $R_j{}^i{}_{kl}$ in terms of R^i_k, C_{ijk} and L_{ijk}.) Plugging (2.73) to (2.71) and (2.72) yields (2.69) and (2.70).   Q.E.D.

The identities (2.69) and (2.70) play an important role in the global Finsler geometry. Later on, we will use them to establish some rigidity theorems.

Chapter 3

Geodesics

In this chapter, we are going to introduce the notion of geodesics and discuss Finsler metrics having the same geodesics as point sets. In particular, we will discuss projectively flat Finsler metrics.

3.1 Sprays

Let M be an n-dimensional manifold and $\pi : TM_o := TM \setminus \{0\} \to M$ be the natural projection. A *spray* G on a manifold M is a special smooth vector field on TM_o in the following form,

$$G = y^i \frac{\partial}{\partial x^i} - 2G^i \frac{\partial}{\partial y^i},$$

where $G^i = G^i(x, y)$ are local functions with the following homogeneity:

$$G^i(x, \lambda y) = \lambda^2 G^i(x, y), \qquad \lambda > 0.$$

A curve $\gamma = \gamma(t)$ in TM_o is called an *integral curve* of G if it satisfies

$$\dot{\gamma} = G_\gamma. \tag{3.1}$$

Let $\gamma(t)$ be an integral curve of G. Then the coordinates $(x^i(t), y^i(t))$ of $\gamma(t)$ satisfy

$$\dot{x}^i(t) = y^i(t), \qquad \dot{y}^i(t) + 2G^i(x(t), y(t)) = 0.$$

Let $\sigma(t) := \pi(\gamma(t))$ be the projection of $\gamma(t)$ under π. Then the coordinates

$(x^i(t))$ of $\sigma(t)$ satisfy

$$\ddot{\sigma}^i(t) + 2G^i\Big(\sigma(t), \dot{\sigma}(t)\Big) = 0. \tag{3.2}$$

Here we identify $\sigma(t)$ and $\dot{\sigma}(t) = \dot{\sigma}^i(t)\frac{\partial}{\partial x^i}|_{\sigma(t)}$ with their coordinates $(\sigma^i(t))$ and $(\dot{\sigma}^i(t))$.

Conversely, given a curve $\sigma = \sigma(t)$ in M, the canonical lift of σ is defined as the curve formed by its tangent vector field,

$$\dot{\sigma}(t) = \dot{\sigma}^i(t)\frac{\partial}{\partial x^i}|_{\sigma(t)}.$$

The coordinates of $\dot{\sigma}(t)$ in TM are $(\sigma^i(t), \dot{\sigma}^i(t))$. It is easy to see that if $\sigma(t)$ satisfies (3.2), then $(x^i(t), y^i(t)) = (\sigma^i(t), \dot{\sigma}^i(t))$ satisfy (3.1). Namely, the canonical lift of σ is an integral curve of G.

A map $\sigma = \sigma(t)$ in M is called a *geodesic* of G if it is a C^∞ curve, and the canonical lift $\gamma(t) := \dot{\sigma}(t)$ is an integral curve of G in TM_o, i.e., it satisfies (3.1). In a standard local coordinate system, the coordinates $(\sigma^i(t))$ of $\sigma(t)$ satisfy (3.2).

Every Finsler metric $F = F(x,y)$ on a manifold M induces a spray $G = y^i\frac{\partial}{\partial x^i} - 2G^i\frac{\partial}{\partial y^i}$ by (2.14) or (2.15), that is,

$$G^i = \frac{1}{4}g^{il}\Big\{[F^2]_{x^k y^l}y^k - [F^2]_{x^l}\Big\}.$$

By the above formula, we can find a formula for the spray coefficients G^i of an (α, β)-metric, which are expressed in terms of the spray coefficients of α and the covariant derivatives of β. Let

$$F = \alpha\,\phi(s) \qquad s := \frac{\beta}{\alpha},$$

where $\phi = \phi(s)$ is a C^∞ function on an interval $(-b_o, b_o)$ satisfying (1.6), $\alpha = \sqrt{a_{ij}(x)y^i y^j}$ is a Riemannian metric on M and $\beta = b_i(x)y^i$ is a 1-form on M. Assume that $\|\beta_x\|_\alpha = \sqrt{a^{ij}(x)b_i(x)b_j(x)} < b_o$ for all $x \in M$. Then F is a Finsler metric.

Let $G^i_\alpha = G^i_\alpha(x,y)$ denote the spray coefficients of α. By (2.13), $G^i_\alpha = \frac{1}{2}\bar{\Gamma}^i_{jk}y^j y^k$, where $\bar{\Gamma}^i_{jk} = \bar{\Gamma}^i_{jk}(x)$ denote the *Christoffel symbols* of α, which are given by

$$\bar{\Gamma}^i_{jk} = \frac{a^{il}}{2}\Big\{\frac{\partial a_{jl}}{\partial x^k} + \frac{\partial a_{kl}}{\partial x^j} - \frac{\partial a_{jk}}{\partial x^l}\Big\}.$$

Let $\theta^i := dx^i$ and $\theta_j{}^i := \bar{\Gamma}^i_{jk}(x)dx^k$. Define $b_{i;j}$ by

$$b_{i;j}\theta^j := db_i - b_j\theta_i{}^j.$$

We have

$$b_{i;j} = \frac{\partial b_i}{\partial x^j} - b_k\bar{\Gamma}^k_{ij}.$$

Let

$$r_{ij} := \frac{1}{2}\Big(b_{i;j} + b_{j;i}\Big), \qquad s_{ij} := \frac{1}{2}\Big(b_{i;j} - b_{j;i}\Big), \tag{3.3}$$

Since $\bar{\Gamma}^k_{ij} = \bar{\Gamma}^k_{ji}$, we have

$$s_{ij} = \frac{1}{2}\Big(\frac{\partial b_i}{\partial x^j} - \frac{\partial b_j}{\partial x^i}\Big).$$

On the other hand, the 1-form $\beta = b_i y^i$, which can be expressed as $\beta = b_i dx^i$, has the following the differential,

$$d\beta = db_j \wedge dx^j = \frac{1}{2}\Big(\frac{\partial b_j}{\partial x^i} - \frac{\partial b_i}{\partial x^j}\Big)dx^i \wedge dx^j.$$

Thus β is closed ($d\beta = 0$) if and only if $s_{ij} = 0$.

Let

$$s^i{}_j := a^{ih}s_{hj}, \qquad s_j := b_i s^i{}_j. \tag{3.4}$$

By (2.13) and using a Maple program (see Section A.3 below), we obtain the following relationship between G^i and G^i_α.

Lemma 3.1.1 *The spray coefficients G^i are related to G^i_α by*

$$G^i = G^i_\alpha + \frac{\alpha\phi'}{\phi - s\phi'}s^i{}_0 + \frac{(\phi - s\phi')\phi' - s\phi\phi''}{2\phi\big((\phi - s\phi') + (b^2 - s^2)\phi''\big)}$$

$$\times\Big\{\frac{-2\alpha\phi'}{\phi - s\phi'}s_0 + r_{00}\Big\}\Big\{\frac{\phi\phi''}{(\phi - s\phi')\phi' - s\phi\phi''}b^i + \frac{y^i}{\alpha}\Big\}, \tag{3.5}$$

where $s = \beta/\alpha$, $s^i{}_0 = s^i{}_j y^j$, $s_0 := s_i y^i$, $r_{00} = r_{ij}y^i y^j$ and $b^2 := a^{ij}b_i b_j$.

A similar formula for the spray coefficients of an (α, β)-metric is given in [51], [68] and [69]. Note that if $\phi'' = 0$, i.e., ϕ is linear in s, then b^i do not occur in (3.5).

Let $F = \alpha + \beta$ be a Randers metric on a manifold M, where $\alpha = \sqrt{a_{ij}y^i y^j}$ and $\beta = b_i y^i$. F can be expressed in the form $F = \alpha\phi(\beta/\alpha)$ where $\phi(s) = 1 + s$. Thus F is a special (α, β)-metric. By Lemma 3.1.1, we obtain the following

Lemma 3.1.2   *For a Randers metric $F = \alpha + \beta$, the relationship between the spray coefficients G^i of F and G_α^i of α is given by*

$$G^i = G_\alpha^i + Py^i + Q^i, \tag{3.6}$$

where

$$P := \frac{e_{00}}{2F} - s_0, \qquad Q^i = \alpha s^i_{\ 0}, \tag{3.7}$$

where $e_{ij} = r_{ij} + b_i s_j + b_j s_i$, $e_{00} := e_{ij}y^i y^j$, $s_0 := s_i y^i$ and $s^i_{\ 0} := s^i_{\ j} y^j$.

Formula (3.6) is given in [2]. By (3.6), one can show that the following three conditions are equivalent for a Randers metric $F = \alpha + \beta$,

  (a) F is a Landsberg metric;
  (b) F is a Berwald metric;
  (c) β is parallel with respect to α, i.e., $b_{i;j} = 0$.

This is a result obtained by several people. See [65], [42], [52], and [95].

As we know that every Randers metric $F = \alpha + \beta$ can be expressed in terms of a Riemannian metric $h = \sqrt{h_{ij}y^i y^j}$ and a vector field $V = V^i \frac{\partial}{\partial x^i}$ by equation (1.39),

$$h\left(x, \frac{y}{F} - V_x\right) = 1.$$

By (1.40) and (1.41), we can express F by

$$F = \frac{\sqrt{\lambda h^2 + V_0}}{\lambda} - \frac{V_0}{\lambda}, \tag{3.8}$$

where $V_0 := V_i y^i$ and $\lambda = 1 - \|V\|_h^2$. Let

$$\phi := \sqrt{1 + s^2} - s$$

and

$$\alpha := \frac{h}{\sqrt{\lambda}}, \qquad \beta := \frac{V_0}{\lambda}.$$

Then F can be expressed as

$$F = \alpha\phi(s), \qquad s := \frac{\beta}{\alpha}.$$

Let $\alpha = \sqrt{a_{ij}y^iy^j}$ and $\beta = b_iy^i$. Define r_{ij}, s_{ij} and s_i as in Lemma 3.1.1. We obtain a formula for the spray coefficients G^i of F which are expressed in terms of the spray coefficients of α and the covariant deri256ives of β with respect to α.

$$G^i = G^i_\alpha - \phi s^i{}_0 + \frac{1}{2(1+b^2)\phi}\Big\{2\phi s_0 + r_{00}\Big\}\Big\{\phi b^i - \frac{y^i}{\alpha}\Big\}. \qquad (3.9)$$

We want to express G^i in terms of the spray coefficients of h and the covariant derivatives of V with respect to h.

First it is easy to verify that

$$1 + b^2 = \frac{1}{\lambda}, \qquad h^2 - 2FV_0 = \lambda F^2. \qquad (3.10)$$

Let

$$\mathcal{R}_{ij} := \frac{1}{2}\Big\{V_{i|j} + V_{j|i}\Big\}, \qquad \mathcal{S}_{ij} := \frac{1}{2}\Big\{V_{i|j} - V_{j|i}\Big\},$$

where "$|$" denotes the covariant differentiation with respect to h. We shall use h^{ij} to lift the lower indices and h_{ij} to lower the upper indices, such as $V^i := h^{ij}V_j$ and $\mathcal{S}^i{}_j := h^{ik}\mathcal{S}_{kj}$, etc. Let

$$\mathcal{R}_j := V^i\mathcal{R}_{ij}, \qquad \mathcal{R} := \mathcal{R}_jV^j, \qquad \mathcal{S}_j := V^i\mathcal{S}_{ij}.$$

Note that $\mathcal{S}_jV^j = 0$. We have

$$\lambda_{|i} = -2(\mathcal{R}_i + \mathcal{S}_i).$$

The spray coefficients G^i_α is related to that of h by

$$G^i_\alpha = G^i_h + \frac{1}{\lambda}(\mathcal{R}_0 + \mathcal{S}_0)y^i - \frac{1}{2\lambda}(\mathcal{R}^i + \mathcal{S}^i)h^2,$$

where $\mathcal{R}_0 := \mathcal{R}_iy^i$ and $\mathcal{S}_0 := \mathcal{S}_iy^i$. Using the above formula for G^i_α, we get

$$r_{00} = \frac{1}{\lambda}\mathcal{R}_{00} + \frac{1}{\lambda^2}\mathcal{R}h^2,$$

$$s^i{}_0 = \mathcal{S}^i{}_0 + \frac{1}{\lambda}\Big\{(\mathcal{R}_0 + \mathcal{S}_0)V^i - (\mathcal{R}^i + \mathcal{S}^i)V_0\Big\},$$

$$s_0 = \frac{1}{\lambda}\mathcal{S}_0 + \frac{1}{\lambda^2}\Big\{(\mathcal{R}_0 + \mathcal{S}_0)(1 - \lambda) - \mathcal{R}V_0\Big\}.$$

By (3.9) and the above formulas, we obtain the following

Lemma 3.1.3   *For a Randers metric F expressed in terms of a Riemannian metric h and a vector field V by (3.8), the spray coefficients G^i of F can be expressed in terms of the spray coefficients G^i_h of h and the covariant derivatives of V with respect to h as follows:*

$$G^i = G^i_h - FS^i{}_0 - \frac{1}{2}F^2(\mathcal{R}^i + \mathcal{S}^i) + \frac{1}{2}\Big\{\frac{y^i}{F} - V^i\Big\}\Big\{2F\mathcal{R}_0 - \mathcal{R}_{00} - F^2\mathcal{R}\Big\}. \quad (3.11)$$

Formula (3.11) is obtained by C. Robles in a different approach [82].

3.2   Shortest Paths

Every Finsler metric F on a manifold M induces a spray G defined in (2.13). Thus the geodesics of G are called the *geodesics* of F. We have the following

Lemma 3.2.1   *If a C^∞ curve $\sigma(t)$ in a Finsler manifold (M, F) is a geodesic, then it has constant speed.*

Proof. It follows from (2.8) and (3.2) that

$$\frac{d}{dt}\Big[F^2\big(\sigma(t), \dot\sigma(t)\big)\Big] = \frac{\partial g_{ij}}{\partial x^k}\dot\sigma^k\dot\sigma^i\dot\sigma^j + 2\frac{\partial g_{ij}}{\partial y^k}\ddot\sigma^k\dot\sigma^i\dot\sigma^j + 2g_{ij}\ddot\sigma^i\dot\sigma^j$$

$$= g_{mj}N^m_i\dot\sigma^i\dot\sigma^j + g_{im}N^m_j\dot\sigma^i\dot\sigma^j - 4g_{ij}G^i\dot\sigma^j = 0.$$

Thus $F(\sigma(t), \dot\sigma(t)) = constant.$ Q.E.D.

A piecewise C^∞ curve C from a point p to another point q is called a *shortest path* if

$$d_F(p, q) = \mathcal{L}_F(C).$$

We shall show the following

Proposition 3.2.2   *For a shortest path in a Finsler manifold, any parametrization with constant speed is a smooth geodesic.*

Proof. Let C be a shortest path from p to q. Parametrize it by $\sigma(t)$, $a \leq t \leq b$, with constant speed $F(\sigma(t), \dot\sigma(t)) = \lambda > 0$. Take an arbitrary piecewise C^∞ vector field $V = V^i(t)\frac{\partial}{\partial x^i}|_{\sigma(t)}$ along σ with $V(a) = 0 = V(b)$. There is a piecewise C^∞ map $H : [a, b] \times (-\varepsilon, \varepsilon) \to M$ such that

$$H(t, 0) = \sigma(t), \qquad \frac{\partial H}{\partial s}(t, 0) = V(t), \qquad a \le t \le b,$$

and

$$H(a, s) = \sigma(a), \qquad H(b, s) = \sigma(b), \qquad |s| < \varepsilon.$$

One may assume that H is C^∞ on each $[t_{i-1}, t_i] \times (-\varepsilon, \varepsilon)$ for some partition $a = t_0 < t_1 < \cdots < t_{k-1} < t_k = b$. Let $\mathcal{L}(s) := \mathcal{L}_F(C_s)$ denote the length of C_s parametrized by $\sigma_s(t) := H(t, s)$.

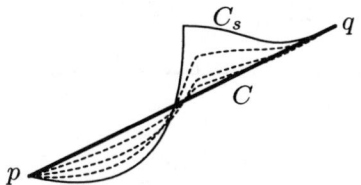

Figure 3.1

By assumption,

$$\mathcal{L}(s) := \mathcal{L}_F(C_s) \ge \mathcal{L}_F(C_0) =: \mathcal{L}(0).$$

Thus $\mathcal{L}'(0) = 0$. To compute the derivative $\mathcal{L}'(0)$, we express $\mathcal{L}(s)$ as

$$\mathcal{L}(s) = \int_a^b F\left(H(t, s), \frac{\partial H}{\partial t}(t, s)\right) dt.$$

Then

$$
\begin{aligned}
\mathcal{L}'(0) &= \int_a^b \frac{1}{2F} \frac{d}{ds}[F^2]\Big|_{s=0} dt \\
&= \int_a^b \frac{1}{2F}\left\{[F^2]_{x^k} V^k + [F^2]_{y^k} \frac{dV^k}{dt}\right\} dt \\
&= \int_a^b \frac{1}{2F}\left\{[F^2]_{x^k} - [F^2]_{x^l y^k}\dot{\sigma}^l - [F^2]_{y^k y^l}\ddot{\sigma}^l\right\} V^k\, dt \\
&\quad + \sum_{i=1}^k \frac{1}{2F}[F^2]_{y^k} V^k\Big|_{t_{i-1}}^{t_i}
\end{aligned}
$$

$$= -\int_a^b \frac{1}{F} g_{jk} \left\{ \ddot{\sigma}^j + 2G^j(\sigma, \dot{\sigma}) \right\} V^k dt$$

$$+ \sum_{i=1}^{k} \frac{1}{F} g_{jk} \dot{\sigma}^j V^k \Big|_{t_{i-1}}^{t_i},$$

where

$$G^i := \frac{1}{4} g^{il}(x, y) \left\{ [F^2]_{x^k y^l}(x, y) y^k - [F^2]_{x^l}(x, y) \right\}.$$

Note that G^i are just the spray coefficients defined in (2.14).

Now, take an arbitrary vector field $V(t) = V^i(t) \frac{\partial}{\partial x^i}\big|_{c(t)}$ along σ with

$$V^i(t) := f(t) \left\{ \ddot{\sigma}^i(t) + 2G^i\big(\sigma(t), \dot{\sigma}(t)\big) \right\},$$

where $f(t)$ is an arbitrary C^∞ function such that $f(t) > 0$ on (t_{i-1}, t_i) and $f(t_i) = 0$ for $i = 0, \cdots, k$.

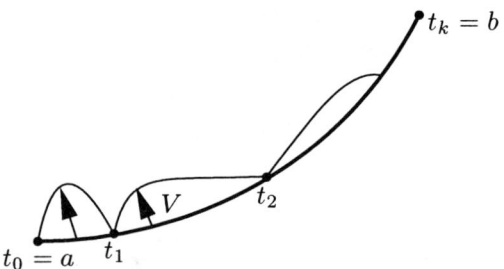

Figure 3.2

From the above formula for $\mathcal{L}'(0)$ and the equation $\mathcal{L}'(0) = 0$, one concludes that σ satisfies (3.2) on each (t_{i-1}, t_i). Then

$$\mathcal{L}'(0) = \sum_{i=1}^{k} \frac{1}{F} g_{jk} \dot{\sigma}^j V^k \Big|_{t_{i-1}}^{t_i}.$$

By choosing a suitable vector field $V(t)$, one immediately sees that $\dot{\sigma}(t_i^+) = \dot{\sigma}(t_i^-)$. Thus $\sigma(t)$ is C^1 at each t_i. Since $\sigma(t)$ satisfies (3.2), one concludes that σ is C^∞ at each t_i. Thus any constant speed parametrization of a shortest path is a geodesic. This proves the above claim. Q.E.D.

Proposition 3.2.3 *Any geodesic $\sigma(t)$ is locally minimizing, namely, at any point t_o in its domain, there is a small number $\varepsilon > 0$ such that the path*

$C = \{\sigma(t) \mid t_o - \varepsilon \le t \le t_o + \varepsilon\}$ *is the shortest path from* $p = \sigma(t_o - \varepsilon)$ *to* $q = \sigma(t_o + \varepsilon)$.

The proof is technical. We omit it here. See [5].

A Finsler metric F on a manifold M is said to be *positively (resp. negatively) complete* if every geodesic $\sigma(t)$ on (a, b) can be extended to a geodesic defined on (a, ∞) (resp. $(-\infty, b)$). F is said to be *complete* if it is both positively complete and negatively complete. There are irreversible Finsler metrics which are only positively complete. For example, the Funk metric on a strongly convex domain in R^n is positively complete, but not complete. While the Klein metric is complete. An important fact is that every closed Finsler manifold is complete. We invite the reader to verify the above facts.

Let (M, F) be a positively complete Finsler manifold and let $x \in M$. There is a natural map $\exp_x : T_x M \to M$ defined as follows using geodesics. For a vector $y \in T_x M$, let $\sigma_y(t)$ denote the geodesic with $\sigma_y(0) = x$ and $\dot{\sigma}_y(0) = y$. Then $\exp_x$ is defined by

$$\exp_x(y) := \sigma_y(1). \tag{3.12}$$

$\exp_x$ is called the *exponential map* at x. By the homogeneity of the spray, one has $\sigma_y(t) = \sigma_{ty}(1)$ for $t > 0$. Thus

$$\sigma_y(t) = \exp_x(ty), \qquad t > 0.$$

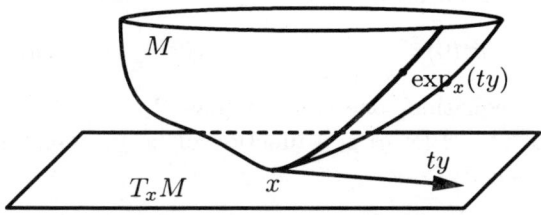

Figure 3.3

The functions $f^i(t, x, y) := \sigma_y^i(t)$ in local coordinates satisfy the following system of second order ordinary equations,

$$\frac{\partial^2 f^i}{\partial t^2}(t, x, y) + 2G^i\left(f(t, x, y), \frac{\partial f}{\partial t}(t, x, y)\right) = 0,$$

with

$$f^i(0, x, y) = x^i, \qquad \frac{\partial f^i}{\partial t}(0, x, y) = y^i.$$

Note that $G^i = G^i(x, y)$ are only C^1 in y at $y = 0$. By the ODE theory, $f^i(t, x, y)$ is C^1 in y at $y = 0$ and C^∞ in y elsewhere. Thus $\exp_x$ is C^∞ on $T_x M \setminus \{0\}$ and C^1 at the origin with $d(\exp_x)|_0 = identity$ [99]. Further, $f^i(t, x, y)$ is C^2 in y at $y = 0$ if and only if $G^i = G^i(x, y)$ are quadratic in $y = y^i \frac{\partial}{\partial x^i}|_x \in T_x M$. Thus, $\exp_x$ is C^2 at the origin of T_xM for all $x \in M$ if and only if F is a Berwald metric. This fact is due to H. Akbar-Zadeh [1].

3.3   Projectively Equivalent Finsler Metrics

In this section, we are going to discuss Finsler metrics having the same geodesics as point sets. In particular, we shall discuss those defined on an open subset in $\mathbf{R}^n$ with straight geodesics.

Two Finsler metrics F and $\bar{F}$ on a manifold M are said to be *projectively equivalent* if they have the same geodesics as point sets, more precisely, for any geodesic $\bar{\sigma}(\bar{t})$ of $\bar{F}$, after an appropriate oriented reparametrization, $\bar{t} = \bar{t}(t)$, the new map $\sigma(t) := \bar{\sigma}(\bar{t}(t))$ is a geodesic of F, and vice versa.

We can characterize projectively equivalent Finsler metrics using the induced sprays. Suppose that F is projectively equivalent to $\bar{F}$. For any $y \in T_x M \setminus \{0\}$, let $\sigma(t)$ be the geodesic of F with $\sigma(0) = x$ and $\dot{\sigma}(0) = y$. There is a transformation $\bar{t} = \bar{t}(t)$ with $\bar{t}(0) = 0$ and $\bar{t}'(0) = 1$ such that $\bar{\sigma}(\bar{t}) := \sigma(t)$ is the geodesic of $\bar{F}$ with $\bar{\sigma}(0) = x$ and $\dot{\bar{\sigma}}(0) = y$. We have

$$2G^i(x, y) = -\ddot{\sigma}^i(0) = -\ddot{\bar{\sigma}}^i(0) - \bar{t}''(0)\dot{\bar{\sigma}}^i(0) = 2\bar{G}^i(x, y) - \bar{t}''(0)y^i.$$

From the above equation, one can see that $P := -\frac{1}{2}\bar{t}''(0)$ depends only on (x, y), hence $P = P(x, y)$ is a function of (x, y). Moreover, P has the following homogeneity

$$P(x, \lambda y) = \lambda P(x, y), \qquad \lambda > 0.$$

Then

$$G^i(x, y) = \bar{G}^i(x, y) + P(x, y)y^i. \tag{3.13}$$

Conversely, if the sprays of Finsler metrics F and $\bar{F}$ are related by (3.13), one can easily show that F and $\bar{F}$ are projectively equivalent.

There is another way to characterize projectively equivalent Finsler metrics. First, let us consider F as a scalar function on TM, and denote by $F_{;k}$ and $F_{.k}$ the horizontal and vertical covariant derivatives of F with respect to $\bar{F}$, which are defined in (2.57), i.e.,

$$dF = F_{;k}\bar{\omega}^k + F_{.k}\bar{\omega}^{n+k},$$

where $\{\bar{\omega}^k, \bar{\omega}^{n+k}\}$ is a local coframe for $T^*(TM_o)$ determined by $\bar{F}$. If $\bar{\omega}_j{}^i$ denote the Chern connection forms with respect to $\{\bar{\omega}^i\}$, then $\bar{\omega}^{n+i} = dy^i + y^j\bar{\omega}_j{}^i$. In a standard local coordinate system (x^i, y^i), $\bar{\omega}^i = dx^i$. Let $\bar{\omega}_j{}^i = \bar{\Gamma}^i_{jk}dx^k$. Then $\bar{\omega}^{n+i} = dy^i + \bar{N}^i_j dx^j$, where $\bar{N}^i_j = \bar{\Gamma}^i_{jk}y^k$. The covariant derivatives $F_{;k}$ and $F_{.k}$ are given by

$$F_{;k} = F_{x^k} - \bar{N}^j_k F_{y^j}, \qquad F_{.k} = F_{y^k}.$$

Let $G^i = G^i(x, y)$ and $\bar{G}^i = \bar{G}^i(x, y)$ denote the spray coefficients of F and $\bar{F}$, respectively, in a common local coordinate system (x^i, y^i). We have

$$\frac{F_{x^k}y^k}{2F} = \frac{F_{;k}y^k}{2F} + \bar{G}^j\frac{F_{y^j}}{F},$$

$$F_{x^k y^l}y^k - F_{x^l} = F_{;k.l}y^k - F_{;l} + 2\bar{G}^j F_{y^j y^l}.$$

Then

$$\frac{F}{2}g^{il}\left\{F_{x^k y^l}y^k - F_{x^l}\right\} = \frac{F}{2}g^{il}\left\{F_{;k.l}y^k - F_{;l}\right\} + \bar{G}^i - \bar{G}^j\frac{F_{y^j}}{F}y^i.$$

Using (2.15) and the above identities, we obtain

$$G^i = \bar{G}^i + Py^i + Q^i, \tag{3.14}$$

where

$$P = \frac{F_{;k}y^k}{2F}, \qquad Q^i = \frac{1}{2}Fg^{il}\left\{F_{;k.l}y^k - F_{;l}\right\}.$$

Note that when $\bar{F}$ is the standard Euclidean metric on $\mathbb{R}^n$, $F_{;k} = F_{x^k}$, then (3.14) is reduced to (2.15).

Theorem 3.3.1 ([80]) *Let F and $\bar{F}$ be Finsler metrics on a manifold M. F is projectively equivalent to $\bar{F}$ if and only if F satisfies the following system,*

$$F_{;k.l}y^k - F_{;l} = 0, \tag{3.15}$$

in which case, their spray coefficients are related by $G^i = \bar{G}^i + Py^i$ *where* $P = \frac{F_{;k}y^k}{2F}$. *Here* $F_{;k}$ *denote the horizontal covariant derivatives of* F *with respect to* $\bar{F}$ *and* $F_{;k\cdot l} = (F_{;k})_{y^l}$.

Proof: We know that F is projectively equivalent to $\bar{F}$ if and only if there is a scalar function $\tilde{P} = \tilde{P}(x,y)$ such that $G^i = \bar{G}^i + \tilde{P}y^i$, which is equivalent to the following equation:

$$Py^i + Q^i = \tilde{P}y^i. \qquad (3.16)$$

Observe that $y^l F_{;k\cdot l} = F_{;k}$. Thus

$$g_{ij}y^j Q^i = \frac{1}{2}Fy^l\left\{F_{;k\cdot l}y^k - F_{;l}\right\} = 0. \qquad (3.17)$$

Assume that F is projectively equivalent to $\bar{F}$. Then (3.16) holds. Contracting (3.16) with $y_i := g_{ij}y^j$ yields

$$P = \tilde{P}.$$

By (3.16) again, one concludes that $Q^i = 0$. This implies (3.15). Conversely, if (3.16) holds, then $Q^i = 0$. It follows from (3.15) that $G^i = \bar{G}^i + Py^i$. Thus F is projectively equivalent to $\bar{F}$. Q.E.D.

Example 3.3.2 ([42]) Let $\bar{F} = \bar{F}(x,y)$ be a Finsler metric and $\beta = b_i(x)y^i$ be a 1-form on a manifold M. Observe that

$$d\beta = db_i \wedge dx^i = \frac{\partial b_i}{\partial x^j}dx^j \wedge dx^i = \frac{1}{2}\left\{\frac{\partial b_j}{\partial x^i} - \frac{\partial b_i}{\partial x^j}\right\}dx^i \wedge dx^j.$$

Consider $F := \bar{F} + \beta$. It follows from (2.58) that $\bar{F}_{;k} = 0$. Here the covariant derivatives are taken with respect to $\bar{F}$. One has

$$F_{;k\cdot l}y^k - F_{;l} = \beta_{;k\cdot l}y^k - \beta_{;l} = \left\{\frac{\partial b_l}{\partial x^k} - \frac{\partial b_k}{\partial x^l}\right\}y^k.$$

By Theorem 3.3.1, one concludes that F is projectively equivalent to $\bar{F}$ if and only if β is closed. In particular, when $\bar{F} = \alpha$ is a Riemannian metric, $F = \alpha + \beta$ is projectively equivalent to α if and only if β is closed.

3.4   Projectively Flat Metrics

Let $\bar{F} = \alpha_0(y)$ be the standard Euclidean norm on $\mathbf{R}^n$. The spray coefficients of $\bar{F}$ vanish, $\bar{G}^i = 0$. Thus for a Finsler metric F on an open subset $\mathcal{U} \subset \mathbf{R}^n$, the geodesics of F are straight lines if and only if the spray coefficients of F are in the following form

$$G^i = Py^i.$$

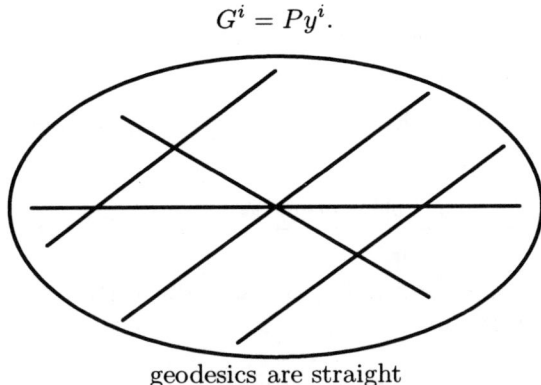

geodesics are straight

Figure 3.4

Straight lines in $\mathcal{U}$ are parametrized by $\sigma(t) = f(t)a + b$, where $a, b \in \mathbf{R}^n$ are constant vectors and $f(t) > 0$ is a positive function. We make the following

Definition 3.4.1   A Finsler metric $F = F(x, y)$ on an open subset $\mathcal{U} \subset \mathbf{R}^n$ is said to be *projectively flat* if all geodesics are straight in $\mathcal{U}$, i.e., $\sigma(t) = f(t)a + b$ for some constant vectors $a, b \in \mathbf{R}^n$. A Finsler metric F on a manifold M is said to be *locally projectively flat* if at any point, there is a local coordinate system (x^i) in which F is projectively flat.

By Theorem 3.3.1, a Finsler metric $F = F(x, y)$ on an open subset $\mathcal{U} \subset \mathbf{R}^n$ is projectively flat if and only if it satisfies the following system of equations,

$$F_{x^k y^l} y^k - F_{x^l} = 0. \tag{3.18}$$

This fact is due to G. Hamel [41]. In this case, $G^i = Py^i$, where $P = P(x, y)$ is given by

$$P = \frac{F_{x^k} y^k}{2F}. \tag{3.19}$$

The scalar function P is called the *projective factor* of F.

Note that (3.18) is a linear equation. Thus if F_1 and F_2 are projectively flat Finsler metrics on $\mathcal{U}$, then $F := aF_1 + bF_2$ is projectively flat on $\mathcal{U}$ as long as F is a Finsler metric. If $F = F(x, y)$ is a projectively flat Finsler metric on $\mathcal{U}$, then its inverse $\bar{F} := F(x, -y)$ is projectively flat on $\mathcal{U}$. Thus the symmetrization $\tilde{F} := \frac{1}{2}\{F + \bar{F}\}$ is also projectively flat on $\mathcal{U}$.

There are many interesting projectively flat Finsler metrics on $\mathbf{R}^n$. Below are some examples.

Example 3.4.2   Consider the following family of Riemannian metrics defined in (1.12):

$$\alpha_\mu := \frac{\sqrt{|y|^2 + \mu(|x|^2|y|^2 - \langle x, y\rangle^2)}}{1 + \mu|x|^2}, \qquad y \in T_x\mathrm{B}^n(r_\mu) \cong \mathbf{R}^n, \qquad (3.20)$$

where $r_\mu = 1/\sqrt{-\mu}$ if $\mu < 0$ and $r_\mu = +\infty$ if $\mu \geq 0$. By a direct computation (see Example 2.3.1), one obtains

$$G^i = -\frac{\mu\langle x, y\rangle}{1 + \mu|x|^2}\, y^i.$$

Thus α_μ is projectively flat. The Beltrami's theorem and Cartan's local classification theorem in Riemannian geometry state that every locally projectively flat Riemannian metric is, up to a scaling, locally isometric to α_μ for some constant μ. However, a projectively flat Riemannian metric may take many different forms. For example, for the family of Riemannian metrics F_a defined in (1.45),

$$F_a = \frac{\sqrt{1 - |a|^2}}{1 + \langle a, x\rangle}\sqrt{|y|^2 - \frac{2\langle a, y\rangle\langle x, y\rangle}{1 + \langle a, x\rangle} - \frac{(1 - |x|^2)\langle a, y\rangle^2}{(1 + \langle a, x\rangle)^2}},$$

where $y \in T_x\mathbf{R}^n \cong \mathbf{R}^n$, the spray coefficients are given by

$$G^i = -\frac{\langle a, y\rangle}{1 + \langle a, x\rangle}y^i.$$

Thus each F_a is projectively flat. In fact, F_a is isometric to α_0. But it is hard to find an isometry between F_a and α_0 which also maps lines to lines.

Example 3.4.3   Let $\mathcal{U} \subset \mathbf{R}^n$ be a strongly convex domain defined by a Minkowski norm $\phi = \phi(y)$ on $\mathbf{R}^n$. Let $\Theta = \Theta(x, y)$ denote the Funk metric on $\mathcal{U}$ defined by (1.35). By (1.38), one can easily verify that Θ satisfies (3.18),

$$\Theta_{x^k y^l} y^k = \Theta_{x^l}.$$

Thus Θ is projectively flat on $\mathcal{U}$. By (3.19) and (1.38), the spray coefficients of Θ are given by

$$G^i = \frac{\Theta_{x^k} y^k}{2\Theta} = \frac{\Theta \Theta_{y^k} y^k}{2\Theta} = \frac{1}{2}\Theta(x,y)y^i.$$

The Finsler metric $\Theta = \Theta(x,y)$ defined by (1.34) is called the *Funk metric* on a strongly convex domain $\mathcal{U}$ in a vector space $\mathbb{R}^n$. Equation (1.38) is the essential property of Θ. Thus we make the following

Definition 3.4.4 A Finsler metric $\Theta = \Theta(x,y)$ on an open subset in $\mathbb{R}^n$ is called a *Funk metric* if it satisfies (1.38).

Example 3.4.5 Let $\Theta = \Theta(x,y)$ be the Funk metric on a strongly convex domain $\mathcal{U} \subset \mathbb{R}^n$ and let

$$H := \frac{1}{2}\Big\{\Theta(x,y) + \Theta(x,-y)\Big\}.$$

Let $\bar{\Theta} := \Theta(x,-y)$. H is called the *Hilbert metric* on $\mathcal{U}$. We are going to show that the Hilbert metric is projectively flat. See also [16], [17] and [77].

By (1.38), $\Theta_{x^k} = \frac{1}{2}[\Theta^2]_{y^k}$, we have

$$\bar{\Theta}_{x^k} = -\frac{1}{2}[\bar{\Theta}]_{y^k}. \tag{3.21}$$

Then

$$\begin{aligned}
H_{x^k y^l} y^k &= \frac{1}{2}\Big\{\Theta_{x^k y^l} + \bar{\Theta}_{x^k y^l}\Big\}y^k \\
&= \frac{1}{4}\Big\{[\Theta^2]_{y^k y^l} - [\bar{\Theta}^2]_{y^k y^l}\Big\}y^k \\
&= \frac{1}{4}\Big\{[\Theta^2]_{y^l} - [\bar{\Theta}^2]_{y^l}\Big\} \\
&= \frac{1}{2}\Big\{\Theta_{x^l} + \bar{\Theta}_{x^l}\Big\} = H_{x^l}.
\end{aligned}$$

That is, H satisfies (3.18). Thus H is projectively flat with $G^i = Py^i$, where P is given by $P = \frac{H_{x^k}y^k}{2H}$. Observe that

$$
\begin{aligned}
P &= \frac{(\Theta_{x^k} + \bar{\Theta}_{x^k})y^k}{2(\Theta + \bar{\Theta})} \\
&= \frac{([\Theta^2]_{y^k} - [\bar{\Theta}^2]_{y^k})y^k}{4(\Theta + \bar{\Theta})} \\
&= \frac{\Theta^2 - \bar{\Theta}^2}{2(\Theta + \bar{\Theta})} \\
&= \frac{1}{2}\left\{\Theta - \bar{\Theta}\right\}.
\end{aligned}
$$

We obtain

$$
P = \frac{1}{2}\left\{\Theta(x, y) - \Theta(x, -y)\right\}.
$$

The Funk metric on the unit ball $B^n(1) \subset R^n$ is given by (1.15).

$$
\Theta = \frac{\sqrt{|y|^2 - (|x|^2|y|^2 - \langle x, y \rangle^2)} + \langle x, y \rangle}{1 - |x|^2}, \qquad y \in T_xB^n \cong R^n.
$$

The corresponding Hilbert metric on $B^n(1)$ is given by

$$
H = \frac{\sqrt{|y|^2 - (|x|^2|y|^2 - \langle x, y \rangle^2)}}{1 - |x|^2}, \qquad y \in T_xB^n \cong R^n.
$$

H is the well-known *Klein metric* on $B^n(1)$.

Example 3.4.6   ([89]) Let $\Theta = \Theta(x, y)$ denote the Funk metric on a strongly convex domain $\mathcal{U} \subset R^n$ (see Example 3.4.3 above). Let $a \in R^n$ be a vector. Set

$$
F := \Theta(x, y) + \frac{\langle a, y \rangle}{1 + \langle a, x \rangle}, \qquad y \in T_x\mathcal{U} \cong R^n. \tag{3.22}
$$

$F = F(x, y)$ is a Finsler metric in a neighborhood of the origin if $|a|$ is sufficiently small. By (1.38), one can easily verify that F satisfies (3.18). Thus it is projectively flat. It follows from (3.19) and (1.38) that $G^i = Py^i$ where

$$
P := \frac{1}{2}\left\{\Theta(x, y) - \frac{\langle a, y \rangle}{1 + \langle a, x \rangle}\right\}.
$$

When $\mathcal{U} = B^n(1)$ is the standard unit ball, $F = F(x, y)$ is a Randers metric given in (1.47),

$$F = \frac{\sqrt{|y|^2 - (|x|^2|y|^2 - \langle x, y\rangle^2)}}{1 - |x|^2} + \frac{\langle x, y\rangle}{1 - |x|^2} + \frac{\langle a, y\rangle}{1 + \langle a, x\rangle}.$$

F is projectively flat with $G^i = Py^i$, where

$$P := \frac{1}{2}\left\{\frac{\sqrt{|y|^2 - (|x|^2|y|^2 - \langle x, y\rangle^2)}}{1 - |x|^2} + \frac{\langle x, y\rangle}{1 - |x|^2} - \frac{\langle a, y\rangle}{1 + \langle a, x\rangle}\right\}.$$

Example 3.4.7 ([43]) Let $\Theta = \Theta(x, y)$ denote the Funk metric on a strongly convex domain $\mathcal{U} \subset R^n$. Define a function $F : T\mathcal{U} \cong \mathcal{U} \times R^n$ by

$$F := \Theta(x, y)\left\{1 + \Theta_{y^m}(x, y)x^m\right\}, \qquad y \in T_x\mathcal{U} \cong R^n.$$

For a point $x \in \mathcal{U}$ sufficiently close to the origin, F_x is a Minkowski norm on $T_x\mathcal{U}$. Thus F is a Finsler metric in a neighborhood of the origin. It follows from (1.38) that

$$F_{x^k}(x, y) = (F\Theta)_{y^k}(x, y).$$

Observe that

$$F_{x^k y^l}(x, y)y^k = (F\Theta)_{y^k y^l}(x, y)y^k = (F\Theta)_{y^l}(x, y) = F_{x^l}(x, y).$$

Thus F satisfies (3.18) and it is projectively flat. The spray coefficients take the form $G^i = Py^i$, where $P = P(x, y)$ is given by (3.19). Using (1.38), one obtains

$$P = \frac{(F\Theta)_{y^k}y^k}{2F} = \frac{2F\Theta}{2F} = \Theta.$$

Thus

$$G^i = \Theta(x, y)\, y^i.$$

When $\mathcal{U} = B^n(1)$ is the standard unit ball in R^n, the Funk metric $\Theta = \Theta(x, y)$ on $B^n(1)$ is given by (1.15) and the corresponding Finsler metric $F = F(x, y)$ is given by (2.56).

It is a natural problem to study locally projectively flat Randers metrics. We have the following

Proposition 3.4.8   *A Randers metric $F = \alpha + \beta$ is locally projectively flat if and only if α is locally projectively flat and β is closed.*

Proof. Suppose that $F = \alpha + \beta$ is locally projectively flat. There is a local coordinate system (x^i) and a scalar function $\tilde{P}$ such that $G^i = \tilde{P}y^i$. By Lemma 3.1.2,

$$G^i = G^i_\alpha + Py^i + Q^i,$$

where $P = e_{00}/(2F) - s_0$ and $Q^i = \alpha s^i{}_0$. We obtain

$$G^i_\alpha + Py^i + Q^i = \tilde{P}y^i. \tag{3.23}$$

Note that

$$\frac{\partial Q^m}{\partial y^m} = \alpha^{-1} y_m s^m{}_0 + \alpha s^m{}_m = 0.$$

Thus

$$\frac{\partial G^m_\alpha}{\partial y^m} + (n+1)P = (n+1)\tilde{P}.$$

By (3.23), one obtains

$$\alpha s^i{}_0 = Q^i = \frac{1}{n+1} \frac{\partial G^m_\alpha}{\partial y^m} \, y^i - G^i_\alpha. \tag{3.24}$$

Note that the right-hand side is quadratic in $y \in T_xM$. Thus both sides are identically zero, that is,

$$s^i{}_0 = 0, \qquad G^i_\alpha = \frac{1}{n+1} \frac{\partial G^m_\alpha}{\partial y^m} \, y^i.$$

The first equation implies that β is closed. The second equation implies that α is projectively flat. The converse is obvious, so the proof is omitted here. See [11] for related discussion.   Q.E.D.

Now we consider a larger class of Finsler metrics which contains Randers metrics.

Example 3.4.9 Let $\alpha = \sqrt{a_{ij}y^i y^j}$ be a Riemannian metric and $\beta = b_i y^i$ be 1-form on a manifold M. Consider an (α, β)-metric in the following form

$$F = \alpha + \varepsilon\beta + k\frac{\beta^2}{\alpha},$$

where ε and k are constants with $k \neq 0$. Then by (3.5), the spray coefficients G^i of F are given by

$$G^i = G^i_\alpha + \frac{(\varepsilon\alpha + 2k\beta)\alpha^2}{\alpha^2 - k\beta^2} s^i{}_0$$
$$+ \frac{\varepsilon\alpha^3 - 3k\varepsilon\alpha\beta^2 - 4k^2\beta^3}{2F\big((1 + 2kb^2)\alpha^2 - 3k\beta^2\big)} \left\{ \frac{-2(\varepsilon\alpha + 2k\beta)\alpha^2}{\alpha^2 - k\beta^2} s_0 + r_{00} \right\}$$
$$\times \left\{ \frac{y^i}{\alpha} + \frac{2k\alpha(\alpha^2 + \varepsilon\alpha\beta + k\beta^2)}{\varepsilon\alpha^3 - 3k\varepsilon\alpha\beta^2 - 4k^2\beta^3} b^i \right\}, \tag{3.25}$$

where G^i_α denote the spray coefficients of α. Assume that β satisfies

$$b_{i|j} = \tau\left\{ (1 + 2kb^2)a_{ij} - 3kb_ib_j \right\}, \tag{3.26}$$

where $\tau = \tau(x)$ is a scalar function on M, then $s^i{}_0 = 0$, $s_0 = 0$ and

$$r_{00} = \tau\left\{ (1 + 2kb^2)\alpha^2 - 3k\beta^2 \right\}.$$

The spray coefficients G^i of F are reduced to

$$G^i = G^i_\alpha + \tau\left\{ \frac{\varepsilon\alpha^3 - 3k\varepsilon\alpha\beta^2 - 4k^2\beta^3}{2(\alpha^2 + \varepsilon\alpha\beta + k\beta^2)} y^i + k\alpha^2 b^i \right\}. \tag{3.27}$$

Further, we assume that G^i_α are in the following form

$$G^i_\alpha = \theta y^i - k\tau\alpha^2 b^i, \tag{3.28}$$

where $\theta = \theta_i y^i$ is a 1-form, then

$$G^i = \left\{ \theta + \tau\frac{\varepsilon\alpha^3 - 3k\varepsilon\alpha\beta^2 - 4k^2\beta^3}{2(\alpha^2 + \varepsilon\alpha\beta + k\beta^2)} \right\} y^i.$$

In this case, F is projectively flat in the coordinate domain. It can be shown that (3.26) and (3.28) are the necessary condition for F being projectively flat when $k \neq 0$.

The next problem is how to find α and β satisfying (3.26) and (3.28). Below is a particular solution.

Let $\alpha = \sqrt{a_{ij}y^i y^j}$ and $\beta = b_i y^i$ be defined by

$$\alpha := \frac{\sqrt{|y|^2 - (|x|^2|y|^2 - \langle x, y \rangle^2)}}{(1 - |x|^2)^2},$$

$$\beta := \frac{\langle x, y \rangle}{(1 - |x|^2)^2}.$$

α is a Riemannian metric and β is a 1-form on the unit ball $B^n(1)$. We have

$$b_{i|j} = \tau\left\{ (1 + 2b^2)a_{ij} - 3b_i b_j \right\}$$

and

$$G^i_\alpha = \tau\left\{ 3\beta y^i - \alpha^2 b^i \right\},$$

where $\tau := 1 - |x|^2$. Consider the following family of (α, β)-metrics,

$$F = \alpha + 2\beta + \frac{\beta^2}{\alpha} = \frac{(\alpha + \beta)^2}{\alpha}. \tag{3.29}$$

By (3.27), the spray coefficients G^i of F are given by

$$G^i = G^i_\alpha + \tau\left\{ \frac{\alpha^3 - 3\alpha\beta^2 - 2\beta^3}{\alpha^2 + 2\alpha\beta + \beta^2} y^i + \alpha^2 b^i \right\}$$

$$= \tau\left\{ 3\beta + \frac{\alpha^3 - 3\alpha\beta^2 - 2\beta^3}{\alpha^2 + 2\alpha\beta + \beta^2} \right\} y^i.$$

Thus F is projectively flat. Note that the metric in (3.29) is the famous Berwald's projectively flat metric with zero flag curvature ([17]). See Example 8.2.8 below for more discussion on this metric.

It is clear that there should be many other $\alpha = \sqrt{a_{ij}y^i y^j}$ and $\beta = b_i y^i$ satisfying (3.27) and (3.28). Some solutions are given in [75].

Chapter 4

Parallel Translations

Parallel translation is a very natural concept in Finsler geometry. In this chapter, we shall first introduce two kinds of parallel translations, then discuss some basic properties of Berwald metrics and Landsberg metrics.

4.1 Parallel Vector Fields

Let (M, F) be a Finsler manifold. Let $c = c(t)$ be a C^∞ curve in M and $U = U^i(t)\frac{\partial}{\partial x^i}|_{c(t)}$ be a vector field along c. Define

$$D_{\dot{c}}U(t) := \left\{ \dot{U}^i(t) + U^j(t)N^i_j\Big(c(t), \dot{c}(t)\Big)\dot{c}^j(t) \right\} \frac{\partial}{\partial x^i}|_{c(t)},$$

where $N^i_j := \Gamma^i_{jk}(x, y)y^k$ are the connection coefficients defined in (2.6). $D_{\dot{c}}U(t)$ is well-defined, independent of local coordinate systems. It is easy to verify that

$$D_{\dot{c}}(U + V)(t) = D_{\dot{c}}U(t) + D_{\dot{c}}V(t), \tag{4.1}$$

$$D_{\dot{c}}(fU)(t) = f'(t)U(t) + f(t)D_{\dot{c}}U(t). \tag{4.2}$$

Since $D_{\dot{c}}U(t)$ linearly depends on $U = U(t)$, $D_{\dot{c}}U(t)$ is called the *linear covariant derivative* of $U(t)$ along c.

A vector field $U = U(t)$ along c is called a *linearly parallel vector field* if it satisfies the equation $D_{\dot{c}}U(t) = 0$, i.e.,

$$\dot{U}^i(t) + U^j(t)N^i_j\Big(c(t), \dot{c}(t)\Big) = 0. \tag{4.3}$$

Clearly, for any t_o in the domain, U linearly depends on the value $U(t_o)$.

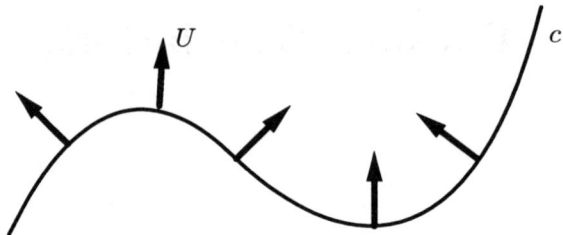

Figure 4.1

Let $\sigma = \sigma(t)$ be a curve in M. Then the tangent vector field $U := \dot{\sigma}(t)$ is a special vector field along σ. Equation (4.3) becomes

$$\ddot{\sigma}(t) + 2G^i\Big(\sigma(t), \dot{\sigma}(t)\Big) = 0. \tag{4.4}$$

Thus a curve σ is a geodesic if and only if its tangent vector field $U = \dot{\sigma}(t)$ is linearly parallel along itself.

Let $\mathcal{T} = T_{ij}(x, y)dx^i \otimes dx^j$ be a tensor on TM_o. Then its covariant derivative is defined by

$$dT_{ij} - T_{kj}\omega_i{}^k - T_{ik}\omega_j{}^k = T_{ij|k}dx^k + T_{ij\cdot k}\delta y^k. \tag{4.5}$$

It follows from (4.5) that

$$T_{ij|k} = \frac{\partial T_{ij}}{\partial x^k}\dot{\sigma}^k - 2G^k\frac{\partial T_{ij}}{\partial y^k} - T_{mj}N_i^m - T_{im}N_j^m.$$

For a non-zero vector $y \in T_xM$, we obtain a multi-linear form $\mathbf{T}_y : T_xM \times T_xM \to \mathbb{R}$ defined by

$$\mathbf{T}_y(u, v) := T_{ij}(x, y)u^iv^j.$$

Let $\sigma = \sigma(t)$ be a geodesic and let $U = U(t)$ and $V = V(t)$ be linearly parallel vector fields along σ. Let

$$\mathbf{T}(t) := \mathbf{T}_{\dot{\sigma}}(U(t), V(t)) = T_{ij}(\sigma(t), \dot{\sigma}(t))U^i(t)V^j(t).$$

Then using (4.3) and (4.4), we have

$$
\begin{aligned}
\mathbf{T}'(t) &= \frac{\partial T_{ij}}{\partial x^k} \dot{\sigma}^k U^i V^j + \frac{\partial T_{ij}}{\partial y^k} \ddot{\sigma}^k U^i V^j + T_{ij} \dot{U}^i V^j + T_{ij} U^i \dot{V}^j \\
&= \frac{\partial T_{ij}}{\partial x^k} \dot{\sigma}^k U^i V^j - 2G^k \frac{\partial T_{ij}}{\partial y^k} U^i V^j - T_{ij} N_m^i U^m V^j - T_{ij} N_m^j U^i V^m \\
&= \left\{ \frac{\partial T_{ij}}{\partial x^k} \dot{\sigma}^k - 2G^k \frac{\partial T_{ij}}{\partial y^k} - T_{mj} N_i^m - T_{im} N_j^m \right\} U^i V^j \\
&= T_{ij|k} \dot{\sigma}^k U^i V^j.
\end{aligned}
$$

For the sake of simplicity, we omitted $(\sigma(t), \dot{\sigma}(t))$ and t in the above identity. Let us re-state it as follows,

$$
\mathbf{T}'(t) = T_{ij|k}(\sigma(t), \dot{\sigma}(t)) \dot{\sigma}^k(t) U^i(t) V^j(t). \tag{4.6}
$$

Lemma 4.1.1 *Let $\sigma = \sigma(t)$ be a geodesic in a Finsler manifold (M, F), and let $U = U(t)$ and $V = V(t)$ be linearly parallel vector fields along σ. Then for the family of induced inner products $\mathbf{g}_{\dot{\sigma}(t)}$ along σ,*

$$
\mathbf{g}_{\dot{\sigma}(t)}\Big(U(t), V(t) \Big) = constant.
$$

Thus the lengths of $U = U(t)$ and $V = V(t)$ and the angle between them with respect to $\mathbf{g}_{\dot{\sigma}}$ are constants along σ.

Proof: It follows from the above argument and the fact that $g_{ij|k} = 0$.
Q.E.D.

There is another notion of covariant derivative along a curve. Let $c = c(t)$ be a piecewise C^∞ curve in a Finsler manifold (M, F). For a vector field $U = U^i(t) \frac{\partial}{\partial x^i}|_{c(t)}$ along c, define the *covariant derivative* $\nabla_{\dot{c}} U(t)$ of U along c by

$$
\nabla_{\dot{c}} U(t) := \left\{ \dot{U}^i(t) + \dot{c}^j(t) N_j^i \Big(c(t), U(t) \Big) \right\} \frac{\partial}{\partial x^i}|_{c(t)}. \tag{4.7}
$$

$\nabla_{\dot{c}} U(t)$ is well-defined, independent of local coordinate systems. However, this ∇ does not satisfy the linearity (4.1) and (4.2). U is said to be *parallel* along c if $\nabla_{\dot{c}} U(t) = 0$, i.e.,

$$
\dot{U}^i + \dot{c}^j(t) N_j^i \Big(c(t), U(t) \Big) = 0. \tag{4.8}
$$

Since $y^j N_j^i = 2G^i$, from (4.8), one can see that a curve $c = c(t)$ is a geodesic if and only if its tangent vector field $U = \dot{c}(t)$ is parallel along itself. Thus we can use either D or ∇ to determine geodesics.

Lemma 4.1.2 *Let $c = c(t)$ be a piecewise C^∞ curve in a Finsler manifold (M, F), and let $U = U^i(t)\frac{\partial}{\partial x^i}|_{c(t)}$ be a parallel vector field along c. Then*

$$F\Big(c(t), U(t)\Big) = constant.$$

Proof: Write

$$F^2\Big(c(t), U(t)\Big) = g_{ij}\Big(c(t), U(t)\Big)U^i(t)U^j(t).$$

Using (2.8), (4.8) and the following fact,

$$C_{ijk}\Big(c(t), U(t)\Big)U^i(t) = 0,$$

one obtains

$$\frac{d}{dt}\Big[F^2\Big(c(t), U(t)\Big)\Big] = \frac{\partial g_{ij}}{\partial x^m}\dot{c}^m U^i U^j + 2g_{ik}U^i\dot{U}^k$$
$$= 2g_{ik}\Gamma^k_{jm}\dot{c}^m U^i U^j - 2g_{ik}N^k_m\dot{c}^m U^i$$
$$= 2g_{ik}N^k_m\dot{c}^m U^i - 2g_{ik}N^k_m\dot{c}^m U^i = 0.$$

Thus $F^2\Big(c(t), U(t)\Big) = constant.$ Q.E.D.

Two Finsler metrics F and $\bar{F}$ on a manifold M are said to be *affinely equivalent* if they have the same geodesics as parametrized curves, that is, if $\sigma = \sigma(t)$ is a geodesic of F, then it is also a geodesic of $\bar{F}$ and vice versa. Let $G^i = G^i(x, y)$ and $\bar{G}^i = \bar{G}^i(x, y)$ denote the spray coefficients of F and $\bar{F}$, respectively, in the same standard local coordinate system (x^i, y^i) in TM. Clearly, F and $\bar{F}$ are affinely equivalent if and only if

$$G^i(x, y) = \bar{G}^i(x, y). \tag{4.9}$$

Thus if a Finsler metric is affinely equivalent to a Riemannian metric, then it must be a Berwald metric.

Assume that $G^i = \bar{G}^i$. By (3.14), we have

$$Py^i + Q^i = 0. \tag{4.10}$$

As shown in (3.17),

$$y_i Q^i = 0,$$

where $y_i := g_{ij} y^j$. Thus

$$0 = y_i (P y^i + Q^i) = P F^2.$$

We obtain $P = 0$. Then by (4.10), we get $Q^i = 0$. It follows from $P = 0$ and $Q^i = 0$ that

$$F_{;k} y^k = 0, \qquad F_{;k \cdot l} y^k - F_{;l} = 0.$$

Observe that

$$(F_{;k} y^k)_{\cdot l} = F_{;k \cdot l} y^k + F_{;l}.$$

Thus

$$F_{;l} = 0. \tag{4.11}$$

Let us remind the reader that $F_{;l}$ denote the horizontal covariant derivatives of F with respect to $\bar{F}$. We obtain the following

Theorem 4.1.3 *Let F and $\bar{F}$ be Finsler metrics on a manifold M. F is affinely equivalent to $\bar{F}$ if and only if F satisfies (4.11).*

Example 4.1.4 Let F be a Minkowski metric on a vector space V. In a standard global coordinate system (x^i, y^i) in $TV \cong V \times V$, $F_{;k} = F_{x^k} = 0$. Thus F is affinely equivalent to the Euclidean metric.

To study affinely equivalent Finsler metrics, one needs the following

Lemma 4.1.5 *Let $(M, \bar{F})$ be a Finsler manifold. If F is another Finsler metric on M such that for any $\bar{F}$-parallel vector field $U = U(t)$ along any curve $c = c(t)$,*

$$F\Big(c(t), U(t)\Big) = constant, \tag{4.12}$$

then F is affinely equivalent to $\bar{F}$.

Proof: For any $\bar{F}$-parallel vector field $U = U(t)$ along c,

$$\frac{d}{dt}\Big[F\big(c, U\big)\Big] = F_{x^j}(c, U)\dot{c}^j + F_{y^i}(c, U)\dot{U}^i$$

$$= F_{x^j}(c, U)\dot{c}^j - F_{y^i}(c, U)N_j^i(c, U)\dot{c}^j$$

$$= \Big\{F_{x^j}(c, U) - F_{y^i}(c, U)N_j^i(c, U)\Big\}\dot{c}^j$$

$$= F_{;j}(c, U)\dot{c}^j. \tag{4.13}$$

Suppose that F is affinely equivalent to $\bar{F}$. Then by Theorem 4.1.3, $F_{;k} = 0$. Thus for any parallel vector field $U = U(t)$ along $c = c(t)$, $F(c(t), U(t)) = constant$. Conversely, suppose that (4.12) holds. For any $x \in M$, any non-zero vectors $y, u \in T_x M$, let $c = c(t)$ be a curve with $c(0) = x, \dot{c}(0) = y$ and $U = U(t)$ be a parallel vector field along c with $U(0) = u$. By assumption,

$$F_{;j}(x, u)y^j = 0.$$

We conclude that $F_{;j} = 0$. By Theorem 4.1.3, F is affinely equivalent to $\bar{F}$.
Q.E.D.

4.2 Parallel Translations

Using parallel vector fields along a curve, one can define parallel translations.

Definition 4.2.1 Let $c = c(t)$, $a \le t \le b$, be a piecewise C^∞ curve from $c(a) = p$ to $c(b) = q$.

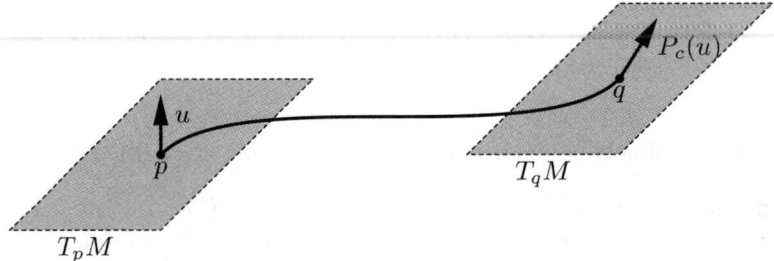

Figure 4.2

Define $P_c : T_p M \to T_q M$ by

$$P_c(u) := U(b), \qquad u \in T_p M,$$

where $U = U(t)$ is the parallel vector field along c with $U(a) = u$. P_c is called the *parallel translation* along c.

By Lemma 4.1.2, the parallel translation P_c is a C^∞ diffeomorphism from $T_pM \setminus \{0\}$ onto $T_qM \setminus \{0\}$, which is positively homogeneous of degree one,

$$P_c(\lambda u) = \lambda P_c(u), \qquad \lambda > 0, \ u \in T_pM.$$

However, P_c is not linear, in general.

Parallel translations can be used to define a group on a Finsler manifold (M, F). For a point $p \in M$, denote by $C(p)$ the set of all piecewise C^∞ closed curves $c : [0, 1] \to M$ starting from $p = c(0)$ and ending at $p = c(1)$. Such a curve c is called a *loop* at p, and the set $C(p)$ is called the *loop space* at p. For loops $c_1, c_2 \in C(p)$, define the product $c_1 * c_2 \in C(p)$ by

$$c_1 * c_2(t) := \begin{cases} c_2(2t) & \text{if } 0 \le t \le \frac{1}{2} \\ c_1(2t - 1) & \text{if } \frac{1}{2} \le t \le 1. \end{cases}$$

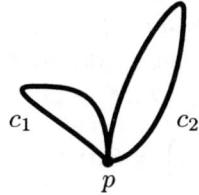

Figure 4.3

Clearly,

$$P_{c_1 * c_2} = P_{c_1} \circ P_{c_2}. \tag{4.14}$$

For a loop $c \in C(p)$, the inverse $c_- = c_-(t)$ of c is also a loop at p defined by

$$c_-(t) := c(1 - t), \qquad 0 \le t \le 1.$$

For an arbitrary parallel vector field $U = U(t)$ along c, and let

$$U_-(t) := U(1 - t), \qquad 0 \le t \le 1.$$

One can verify that $U_- = U_-(t)$ is parallel along c_- with $U_-(0) = U(1)$ and $U_-(1) = U(0)$. Thus

$$P_{c_-} \circ P_c = P_{c_- * c} = identity = P_{c * c_-} = P_c \circ P_{c_-}, \qquad (4.15)$$

Let

$$H_p := \left\{ P_c : T_p M \to T_p M, \ \Big| \ c \in C(p) \right\}.$$

By (4.14) and (4.15), one can see that H_p with the multiplication " $\circ$ " is a group. H_p is called the *holonomy group* at p.

The following proposition shows how to use the holonomy group to construct Finsler metrics which are affinely equivalent to a given Finsler metric.

Proposition 4.2.2 *Let $(M, \bar{F})$ be a Finsler manifold. Let H_p denote the holonomy group of $\bar{F}$ at a point $p \in M$. If F_p is a H_p-invariant Minkowski norm on $T_p M$, then F_p can be extended to a Finsler metric F on M by parallel translations such that F is affinely equivalent to $\bar{F}$.*

Proof: For an arbitrary point $q \in M$, let c be a piecewise C^∞ curve issuing from $c(0) = p$ to $c(1) = q$. Let P_c denote the parallel translation along c with respect to $\bar{F}$. Define $F_q : T_q M \to [0, \infty)$ by

$$F_q \Big(P_c(v) \Big) = F_p(v), \qquad v \in T_p M.$$

If σ is another piecewise C^∞ curve issuing from $\sigma(0) = p$ to $\sigma(1) = q$. Then $\tau := c^{-1} * \sigma$ is a loop at p. It follows from (4.15) that

$$P_\sigma = P_{c * c^{-1}} \circ P_\sigma = P_{c * c^{-1} * \sigma} = P_c \circ P_\tau.$$

Since F_p is H_p-invariant, one obtains

$$F_q \Big(P_\sigma(v) \Big) = F_q \Big(P_c \circ P_\tau(v) \Big) = F_p \Big(P_\tau(v) \Big) = F_p(v) = F_q \Big(P_c(v) \Big).$$

Thus F_q is well-defined. From the above construction, one can see that for any $\bar{F}$-parallel vector field $U = U(t)$ along any curve $c = c(t)$,

$$F \Big(c(t), U(t) \Big) = constant.$$

By Lemma 4.1.5, F is affinely equivalent to $\bar{F}$.   Q.E.D.

4.3 Berwald Metrics

Recall that a Finsler metric F on a manifold M is a Berwald metric if in any standard local coordinate system (x^i, y^i) in TM, the spray coefficients $G^i = \frac{1}{2}\Gamma^i_{jk}(x)y^j y^k$ are quadratic in $y \in T_x M$ for all $x \in M$. Riemannian metrics and Minkowski metrics are trivial Berwald metrics. There are many non-Riemannian Berwald metrics.

Example 4.3.1 Let $f = f(s, t) \geq 0$, $s \geq 0, t \geq 0$, be a C^∞ function satisfying (1.18). Let (M_i, α_i), $i = 1, 2$, be Riemannian manifolds and $M = M_1 \times M_2$. Let

$$F(x, y) := \sqrt{f\Big([\alpha_1(x_1, y_1)]^2, \ [\alpha_2(x_2, y_2)]^2\Big)},$$

where $x = (x_1, x_2) \in M$ and $y = y_1 \oplus y_2 \in T_x M \cong T_{x_1} M_1 \oplus T_{x_2} M_2$. According to Example 1.2.5, F is a Finsler metric if and only if f satisfies (1.20) and (1.21). Let

$$\bar{F}(x, y) := \sqrt{[\alpha_1(x_1, y_1)]^2 + [\alpha_2(x_2, y_2)]^2}.$$

$\bar{F}$ is the standard product of α_1 and α_2. By a direct computation, one knows that the spray coefficients of $\bar{F}$ are split as the direct sum of the spray coefficients of α_1 and α_2, that is,

$$\bar{G}^a(x, y) = \bar{G}^a(x_1, y_1), \qquad \bar{G}^\alpha(x, y) = \bar{G}^\alpha(x_2, y_2). \tag{4.16}$$

Let $c = (c_1(t), c_2(t))$ be a curve in M and $U = U_1(t) \oplus U_2(t)$ be a parallel vector field along c with respect to $\bar{F}$. By (4.16), one can see that each U_i must be a parallel vector field along c_i with respect to α_i, $i = 1, 2$. Thus

$$\alpha_i\Big(c_i(t), U_i(t)\Big) = constant.$$

This implies that

$$F\Big(c(t), U(t)\Big) = constant.$$

By Lemma 4.1.5, one concludes that F is affinely equivalent to $\bar{F}$. The spray coefficients G^i of F are equal to that of $\bar{F}$. Since $\bar{F}$ is a Riemannian metric, $\bar{G}^i = \frac{1}{2}\bar{\Gamma}^i_{jk}(x)y^j y^k$ are quadratic in y, then so are G^i. This shows that F is a Berwald metric.

Let (M, F) be a Berwald manifold. In any local coordinate system,

$$u^j N_j^i(x, y) = u^j \Gamma_{jk}^i(x) y^k = y^j \Gamma_{jk}^i(x) u^k = y^j N_j^i(x, u).$$

Thus for any curve $c = c(t)$ and any vector field $U = U(t)$ along c,

$$D_{\dot{c}} U(t) = \nabla_{\dot{c}} U(t).$$

Therefore, any parallel vector field along a curve linearly depends on its initial value. By Lemma 4.1.2, one immediately obtains the following

Proposition 4.3.2 ([47]) *Let (M, F) be a Berwald manifold. For any piecewise C^∞ curve $c(t)$ from p to q in M, the parallel translation P_c is a linear isometry between $(T_p M, F_p)$ and $(T_q M, F_q)$.*

We know that if a Finsler metric is affinely equivalent to a Riemannian metric, then it is a Berwald metric. Lemma 4.1.5 gives us an idea to construct Berwald metrics from a Riemannian metric. In fact, every Berwald metric can be constructed from a Riemannian metric in this way.

Proposition 4.3.3 ([97]) *A Finsler metric F on a manifold M is a Berwald metric if and only if it is affinely equivalent to a Riemannian metric α. In this case, F and α have the same holonomy group H_p at any point $p \in M$.*

Proof. Assume that F is a Berwald metric. We are going to construct a Riemannian metric α that is affinely equivalent to F. Let D be the Levi-Civita connection of F. Let H_p denote the holonomy group of D. H_p acts on $T_p M$ leaving the indicatrix S_p of F_p invariant. Let G_p denote the subgroup of all linear transformations $\gamma : T_p M \to T_p M$ that leaves S_p invariant. G_p is a compact Lie group and H_p is a subgroup of G_p [98]. In general, H_p is non-compact. Fix a non-zero vector $y_o \in T_p M$ and let $\mathbf{g}_{y_o}$ denote the induced inner product on $T_p M$. Define a Euclidean norm α_p on $T_p M$ by

$$\alpha_p(v) := \frac{\lambda}{\mu(G_p)} \int_{G_p} \sqrt{\mathbf{g}_{y_o}(\gamma v, \gamma v)} \, d\mu(\gamma), \qquad v \in T_p M,$$

where $d\mu$ is the bi-invariant Haar measure on G_p. The constant λ is chosen so that $\alpha_p(y_o) = F_p(y_o)$. From the definition, one can see that α_p is G_p-invariant, hence H_p-invariant. For any point $q \in M$, let $c = c(t)$, $0 \leq t \leq 1$, be a piecewise C^∞ curve issuing from $c(0) = p$ to $c(1) = q$. Define a Euclidean norm α_q in $T_q M$ by

$$\alpha_q(v) := \alpha_p\Big(V(0)\Big), \qquad v \in T_q M,$$

where $V = V(t)$ is the parallel vector field along c with $V(1) = v$. Since α_p is H_p-invariant, α_q is well-defined. One obtains a Riemannian metric $\alpha(q, v) := \alpha_q(v)$, $q \in M$. From the above construction, one can see that for any C^∞ curve c and any F-parallel vector field $U = U(t)$ along c,

$$\alpha\Big(c(t), U(t)\Big) = constant.$$

By Lemma 4.1.5, one concludes that α is affinely equivalent to F. Q.E.D.

According to Proposition 4.3.3, every Berwald metric is affinely equivalent to a Riemannian metric. This observation leads to the classification of Berwald metrics.

Theorem 4.3.4 (Local Structure Theorem [97]) *Let (M, F) be a Berwald manifold. Then (M, F) can be locally decomposed to a product of locally Minkowski manifolds, Riemannian manifolds and locally irreducible locally symmetric Berwald manifolds of rank $r \geq 2$.*

Since any two-dimensional Berwald manifold does not contain any locally irreducible locally symmetric Berwald manifold of rank $r \geq 2$, one obtains the following

Corollary 4.3.5 ([97]) *Any two-dimensional Berwald manifold is either locally Minkowskian or Riemannian.*

4.4 Landsberg Metrics

Recall that a Finsler metric on a manifold is called a Landsberg metric if the Landsberg tensor vanishes, $\mathcal{L} = 0$. See Definition 2.1.2 above. By Proposition 2.1.3, every Berwald metric is a Landsberg metric. But it is still an open problem whether or not there is a Landsberg metric which is not Berwaldian.

Landsberg manifolds have some nice properties. For example, all slit tangent spaces with the induced Riemannian metric are isometric in a canonical way. This fact will be proved below.

Let (M, F) be a Finsler manifold. Take a standard local coordinate system (x^i, y^i) in TM. At each point $x \in M$, $F_x = F|_{T_x M}$ induces a Riemannian metric $\hat{\mathbf{g}}_x$ on $T_x M \setminus \{0\}$,

$$\hat{\mathbf{g}}_x = g_{ij} dy^i \otimes dy^j,$$

where $g_{ij} := \frac{1}{2}[F^2]_{y^i y^j}(x, y)$ and $\{dy^i\}$ is the global natural co-frame on the manifold $T_x M$ corresponding to the basis $\{\frac{\partial}{\partial x^i}|_x\}$ for $T_x M$.

Proposition 4.4.1   ([48]) *Let (M, F) be a Landsberg manifold. Then for any piecewise C^∞ curve c from p to q, the parallel translation P_c along c preserves the induced Riemannian metrics on the slit tangent spaces, i.e., $P_c : (T_p M \setminus \{0\}, \hat{\mathbf{g}}_p) \to (T_q M \setminus \{0\}, \hat{\mathbf{g}}_q)$ is an isometry.*

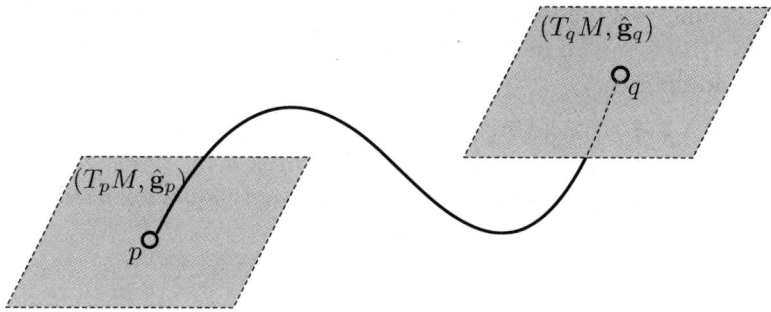

Figure 4.4

Proof. Without loss of generality, one may assume that the curve $c = c(t), 0 \leq t \leq 1$, is embedded in an open domain $\mathcal{U}$ on which there is a non-zero vector field $X = X^i \frac{\partial}{\partial x^i} \in C^\infty(TM)$ such that

$$\dot{c}(t) = X_{c(t)}.$$

Namely, X is an extension of the tangent vector field $\dot{c} = \dot{c}(t)$ along c in a neighborhood $\mathcal{U}$ of c. The horizontal lift $\bar{X}$ of X to TM_o is defined by

$$\bar{X}_{(x,y)} := X^k(x) \left\{ \frac{\partial}{\partial x^k} - N_k^i(x, y) \frac{\partial}{\partial y^i} \right\}_{|(x,y)}.$$

Let H_t and $\bar{H}_t$ denote the flows of X and $\bar{X}$ in $\mathcal{U}$ and $\pi^{-1}\mathcal{U}$ respectively,

$$\frac{dH_t}{dt}(x) = X_{H_t(x)}, \qquad H_0(x) = x, \tag{4.17}$$

$$\frac{d\bar{H}_t}{dt}(x, y) = \bar{X}_{\bar{H}_t(x,y)}, \qquad \bar{H}_0(x, y) = (x, y). \tag{4.18}$$

By (4.17), $H_t(p) = c(t)$. It is easy to verify that

$$\pi \circ \bar{H}_t = H_t.$$

For a vector $y \in T_pM$, let $U(t) := \bar{H}_t(p, y)$. Observe that

$$\pi(U(t)) = \pi \circ \bar{H}_t(p, y) = H_t(p) = c(t).$$

Thus $U(t) = U^i(t) \frac{\partial}{\partial x^i}|_{c(t)}$ is a vector field along c. The local coordinates of $\bar{H}_t(p, y) = U(t)$ in TM are $(c^i(t), U^i(t))$. By (4.18),

$$\frac{d\bar{H}_t}{dt}(p, y) = \bar{X}_{U(t)}.$$

In local coordinates,

$$\dot{c}^i(t) \frac{\partial}{\partial x^i}|_{U(t)} + \dot{U}^i(t) \frac{\partial}{\partial y^i}|_{U(t)} = \dot{c}^k(t) \left\{ \frac{\partial}{\partial x^k} - N_k^i(c(t), U(t)) \frac{\partial}{\partial y^i} \right\}|_{U(t)}.$$

Thus

$$\dot{U}^i(t) + \dot{c}^k(t) N_k^i(c(t), U(t)) = 0.$$

That is, $U = U(t)$ is a parallel vector field along c. This implies that for any $r > 0$, the map

$$P_r := \bar{H}_r(p, \cdot) : T_pM \to T_{c(r)}M$$

is the parallel translation along $c_r = c(t)$, $0 \le t \le r$. In particular, we have

$$P_t(\dot{c}(0)) = \bar{H}_t(p, \dot{c}(0)) = \dot{c}(t).$$

Let $\xi \in T_pM \setminus \{0\}$ and $\eta : (-\varepsilon, \varepsilon) \to T_pM$ be a C^∞ curve with $\eta(0) = \dot{c}(0)$ and $\dot{\eta}(0) = \xi$. Let

$$\bar{H}(t, s) := \bar{H}_t \Big(p, \eta(s) \Big).$$

We have

$$\bar{H}(t, 0) = \bar{H}_t(p, \dot{c}(0)) = \dot{c}(t).$$

The map $\bar{H}$ is a variation of $\dot{c} = \dot{c}(t)$ in TM_o. Put

$$\bar{X}(t, s) := \frac{\partial \bar{H}}{\partial t}(t, s), \qquad V(t, s) := \frac{\partial \bar{H}}{\partial s}(t, s).$$

Since $\bar{X} = \bar{X}(s,t)$ is horizontal and $V = V(s,t)$ is vertical, one has

$$\omega^i(V) = 0, \qquad \omega^{n+i}(\bar{X}) = 0,$$

where $\omega^i = dx^i$ and $\omega^{n+i} = \delta y^i := dy^i + N_j^i dx^j$. Observe that

$$\begin{aligned}
\bar{X}[\omega^{n+i}(V)] &= d\omega^{n+i}(\bar{X}, V) + V\omega^{n+i}(\bar{X}) + \omega^{n+i}([\bar{X}, V]) \\
&= d\omega^{n+i}(\bar{X}, V) \\
&= \left\{\omega^{n+l} \wedge \omega_l{}^i + \Omega^i\right\}(\bar{X}, V) \\
&= \left\{-\omega_l{}^i(\bar{X}) - L^i{}_{kl}\omega^k(\bar{X})\right\}\omega^{n+l}(V).
\end{aligned}$$

Let $\hat{\mathbf{g}} := g_{ij}\omega^{n+i} \otimes \omega^{n+j}$. $\hat{\mathbf{g}}$ is a tensor on TM_o and its pull-back onto $T_x M \setminus \{0\}$ by the natural embedding is the Riemannian metric $\hat{\mathbf{g}}_x$. Observe that

$$\begin{aligned}
\frac{d}{dt}\left[P_t^*(\hat{\mathbf{g}}_{c(t)})\right](\xi, \xi) &= \bar{X}\left[\hat{\mathbf{g}}(V, V)\right] \\
&= \bar{X}\left[g_{ij}\omega^{n+i}(V)\omega^{n+j}(V)\right] \\
&= \bar{X}[g_{ij}]\omega^{n+i}(V)\omega^{n+j}(V) + 2g_{ij}\bar{X}[\omega^{n+i}(V)]\omega^{n+j}(V) \\
&= 2g_{im}\omega^m_j(\bar{X})\omega^{n+i}(V)\omega^{n+j}(V) \\
&\quad -2g_{ij}\omega^i{}_l(\bar{X})\omega^{n+l}(V)\omega^{n+j}(V) \\
&\quad -2g_{ij}L^i{}_{kl}\omega^k(\bar{X})\omega^{n+l}(V)\omega^{n+j}(V) \\
&= -2g_{ij}L^i{}_{kl}\omega^k(\bar{X})\omega^{n+l}(V)\omega^{n+j}(V).
\end{aligned}$$

By assumption, the Landsberg curvature vanishes, $\mathcal{L} = 0$, one concludes that

$$\frac{d}{dt}\left[P_t^*(\hat{\mathbf{g}}_{c(t)})\right](\xi, \xi) = 0.$$

This implies that $P_t : \left(T_p M, \hat{\mathbf{g}}_p\right) \rightarrow \left(T_{c(t)}, \hat{\mathbf{g}}_{c(t)}\right)$ is an isometry for any $0 \le t \le 1$.   Q.E.D.

According to Proposition 4.4.1, if (M, F) is a Landsberg manifold, then for any loop c at $p \in M$, the parallel translation P_c is an isometry of the Riemannian tangent space $(T_p M \setminus \{0\}, \hat{\mathbf{g}}_p)$. Thus H_p is a subgroup of the isometry group G of $(T_p M \setminus \{0\}, \hat{\mathbf{g}}_p)$. The isometry group G is a Lie group [56]. Further, H_p is closed in G. Thus H_p is also a Lie group.

Proposition 4.4.2   ([59], [60]) *For any Landsberg manifold (M, F), the holonomy group H_p at any point $p \in M$ is a Lie group.*

A natural question is whether or not the holonomy group of any Finsler manifold is a Lie group. Is there a Finsler manifold whose holonomy group is not the holonomy group of any Riemannian manifold? These problems remain open.

Chapter 5

S-Curvature

On a Finsler manifold (M, F), the indicatrix of $F_x := F|_{T_xM}$ on the tangent space T_xM can be viewed as the *infinitesimal color pattern* at x. Thus (M, F) is quite "colorful". If F is a Berwald metric, by Proposition 4.3.2, all tangent spaces (T_xM, F_x) are linearly isometric to each other, namely, the infinitesimal color patterns are all the same over the manifold. Intuitively, Berwald manifolds have homogeneous "color". In particular, Riemannian manifolds are entirely "white". It is natural to seek for some quantities which measure the changes in "color" over the Finsler manifold. The Cartan torsion determines the shape of the indicatrix at a point, and the Landsberg tensor is a quantity which measures the changes of indicatrices. The Landsberg tensor is defined on the slit tangent bundle using the Chern connection. In this chapter, we will introduce two new quantities, the distortion and the S-curvature. The distortion is defined on each Minkowski tangent space, which measures the non-Euclidean property of the Minkowski norm, while the S-curvature is the rate of change of the distortion along geodesics. These two quantities are closely related to other quantities.

5.1 Distortion and S-Curvature

The indicatrix of a Minkowski norm on a vector space can be viewed as a color pattern on the vector space. The (mean) Cartan torsion is the (average) rate of tangential change of the color pattern along the indicatrix. Besides the (mean) Cartan torsion, there is another important quantity which also measures certain geometric properties of the indicatrix of a Minkowski norm.

Let V be an n-dimensional vector space and let $F = F(y)$ be a Minkowski norm on V. Fix a basis $\{\mathbf{b}_i\}$ for V and let

$$\sigma_F := \frac{\text{Vol}(B^n)}{\text{Vol}\{(y^i) \in \mathbb{R}^n \mid F(y^i \mathbf{b}_i) < 1\}}.$$

Using σ_F, we define an n-form on V, up to a sign, by

$$dV_F := \sigma_F \theta^1 \wedge \cdots \wedge \theta^n,$$

where $\{\theta^i\}$ is a basis for V^*, dual to $\{\mathbf{b}_i\}$. Clearly, σ_F depends on the choice of a particular basis, while dV_F does not. dV_F is well-defined.

If $F = \sqrt{g_{ij} y^i y^j}$, where $y = y^i \mathbf{b}_i$, is a Euclidean norm, then

$$\text{Vol}\left\{(y^i) \in \mathbb{R}^n \mid F(y^i \mathbf{b}_i) < 1\right\} = \frac{\text{Vol}(B^n)}{\sqrt{\det(g_{ij})}}.$$

Thus

$$\sigma_F = \sqrt{\det(g_{ij})}.$$

Then

$$dV_F = \sqrt{\det(g_{ij})} \theta^1 \wedge \cdots \wedge \theta^n$$

is the Euclidean volume form on V.

If F is a Minkowski norm, $g_{ij} := \frac{1}{2}[F^2]_{y^i y^j}(y)$ depend on y and $\sigma_F \neq \sqrt{\det(g_{ij}(y))}$, in general. Define

$$\tau := \ln \frac{\sqrt{\det\left(g_{ij}(y)\right)}}{\sigma_F}. \tag{5.1}$$

Both σ_F and $\sqrt{\det(g_{ij}(y))}$ are transformed in the same way as the basis $\{\mathbf{b}_i\}$ changes. Thus $\tau = \tau(y)$ is well-defined, which is called the *distortion* of F [85], [87]. Observe that

$$\tau_{y^i} = \frac{\partial}{\partial y^i}\left[\ln \sqrt{\det\left(g_{jk}(y)\right)}\right] = \frac{1}{2} g^{jk} \frac{\partial g_{jk}}{\partial y^i} = g^{jk} C_{ijk},$$

where $C_{ijk} = \frac{1}{2}\frac{\partial g_{jk}}{\partial y^i}$. Recall that the mean Cartan tensor $\mathbf{I} = I_i dx^i$ is given by $I_i = g^{jk} C_{ijk}$. Thus

$$I_i = \tau_{y^i}. \tag{5.2}$$

By Deicke's theorem (Theorem 1.5.1), one concludes that F is Euclidean if and only if $\tau = constant$, in which case, $\tau = 0$. Therefore, a Minkowski norm is Euclidean if and only if $\tau = 0$. We summarize the above arguments in the following lemma.

Lemma 5.1.1 *For a Minkowski norm F on a vector space* V, *the following conditions are equivalent: (a) F is Euclidean, (b) $\tau = constant$ and (c) $\tau = 0$.*

Now we consider Finsler metrics. Let F be a Finsler metric on a manifold M. Since the distortion is defined for the Minkowski norm F_x on every tangent space $T_x M$, we obtain a scalar function $\tau = \tau(x, y)$ on $TM \setminus \{0\}$. We call it the *distortion* of F. By Deicke's theorem, F is Riemannian if and only if $\tau = 0$. Thus the distortion characterizes Riemannian metrics among Finsler metrics.

It is natural to study the rate of change of the distortion along geodesics. For a vector $y \in T_x M \setminus \{0\}$, let $\sigma = \sigma(t)$ be the geodesic with $\sigma(0) = x$ and $\dot\sigma(0) = y$. Set

$$\mathbf{S}(x, y) := \frac{d}{dt}\Big[\tau\big(\sigma(t), \dot\sigma(t)\big)\Big]\big|_{t=0}. \tag{5.3}$$

$\mathbf{S} = \mathbf{S}(x, y)$ is positively y-homogeneous of degree one,

$$\mathbf{S}(x, \lambda y) = \lambda \mathbf{S}(x, y), \qquad \lambda > 0.$$

$\mathbf{S}$ is called the *S-curvature*.

In a standard local coordinate system (x^i, y^i), let $dV_F = \sigma_F(x) dx^1 \wedge \cdots \wedge dx^n$ denote the volume form and $G^i = G^i(x, y)$ denote the spray coefficients of F. It follows from (2.12) that

$$\frac{\partial G^m}{\partial y^m} = \frac{1}{2} g^{ml} \frac{\partial g_{ml}}{\partial x^i} y^i - 2 I_i G^i.$$

Then

$$\begin{aligned}
\mathbf{S} &= y^i \frac{\partial \tau}{\partial x^i} - 2 \frac{\partial \tau}{\partial y^i} G^i \\
&= \frac{1}{2} g^{ml} \frac{\partial g_{ml}}{\partial x^i} y^i - 2 I_i G^i - y^m \frac{\partial}{\partial x^m}\Big(\ln \sigma_F(x)\Big) \\
&= \frac{\partial G^m}{\partial y^m}(x, y) - y^m \frac{\partial}{\partial x^m}\Big(\ln \sigma_F(x)\Big).
\end{aligned} \tag{5.4}$$

There is another quantity associated with the S-curvature. Let

$$E_{ij} := \frac{1}{2}\mathbf{S}_{y^i y^j}(x, y) = \frac{1}{2}\frac{\partial^2}{\partial y^i \partial y^j}\left[\frac{\partial G^m}{\partial y^m}\right](x, y). \tag{5.5}$$

Then $\mathcal{E} := E_{ij}dx^i \otimes dx^j$ is a tensor on TM_o. We call it the *E-tensor*. The E-tensor can be viewed as a family of symmetric forms $\mathbf{E}_y : T_x M \times T_x M \to \mathrm{R}$ defined by

$$\mathbf{E}_y(u, v) := E_{ij}(x, y)u^i v^j,$$

where $u = u^i \frac{\partial}{\partial x^i}|_x, v = v^j \frac{\partial}{\partial x^j}|_x \in T_x M$. Then $\mathbf{E} := \{\mathbf{E}_y \mid y \in TM \setminus \{0\}\}$ is called the *E-curvature* or the *mean Berwald curvature*.

From the definition, it is clear that

$$\mathbf{E}_y(y, v) = \mathbf{E}(u, y) = 0,$$

where $u, v \in T_x M$, since $\mathbf{S}$ is positively y-homogeneous of degree one. Clearly, if $\mathbf{S} = 0$, then $\mathbf{E} = 0$. The converse might not be true. But so far, no counter-example has been found yet.

Proposition 4.3.2 tells us that every Berwald manifold is modeled on a single Minkowski space. Moreover, the geometry of tangent spaces does not change along geodesics. This observation leads to the following

Proposition 5.1.2   ([85]) *For any Berwald metric, the S-curvature vanishes,* $\mathbf{S} = 0$.

Proof: Fix an arbitrary point $(x, y) \in TM_o$ and let $\sigma = \sigma(t)$ be an arbitrary geodesic with $\sigma(0) = x$ and $\dot{\sigma}(0) = y$. Let $\{\mathbf{b}_i(t)\}$ be a linearly parallel frame along σ, namely, each $\mathbf{b}_i(t)$ is linearly parallel along σ. Let

$$g_{ij}(t) := \mathbf{g}_{\dot{\sigma}(t)}\Big(\mathbf{b}_i(t), \mathbf{b}_j(t)\Big).$$

By Lemma 4.1.1, $g_{ij}(t) = constant$. Thus $\det\Big(g_{ij}(t)\Big) = constant$. On the other hand, for any $(y^i) \in \mathrm{R}^n$, the vector field $U = y^i \mathbf{b}_i(t)$ is linearly parallel along σ. By Lemma 4.1.2,

$$F\Big(\sigma(t), y^i \mathbf{b}_i(t)\Big) = constant.$$

Thus the following convex subset $\mathcal{U}_t \subset \mathrm{R}^n$ is independent of t,

$$\mathcal{U}_t := \Big\{(y^i) \in \mathrm{R}^n \mid F\Big(\sigma(t), y^i \mathbf{b}_i(t)\Big) < 1\Big\}.$$

This implies that the coefficient of the volume form dV_F is a constant,

$$\sigma_F(\sigma(t)) = \frac{\text{Vol}(\text{B}^n(1))}{\text{Vol}(\mathcal{U}_t)} = constant.$$

Therefore, the distortion must be a constant along σ, i.e.,

$$\tau\Big(\sigma(t), \dot{\sigma}(t)\Big) = \ln \frac{\sqrt{\det(g_{ij}(t))}}{\sigma_F(\sigma(t))} = constant.$$

Thus $\mathbf{S} = 0$ by (5.3).   Q.E.D.

A Finsler metric F on an n-dimensional manifold M is said to have *almost isotropic S-curvature* if there is a scalar function $c = c(x)$ on M such that

$$\mathbf{S} = (n+1)\Big\{cF + \eta\Big\}, \tag{5.6}$$

where $\eta = \eta_i(x)y^i$ is a closed 1-form. F is said to have *isotropic S-curvature* if $\eta = 0$. F is said to have *constant S-curvature* if $\eta = 0$ and $c = constant$. Similarly, F is said to have *isotropic E-curvature* if there is a scalar function $c = c(x)$ on M such that

$$\mathbf{E} = \frac{1}{2}(n+1)cF^{-1}\mathbf{h}. \tag{5.7}$$

Here $\mathbf{h}$ is a family of bilinear forms $\mathbf{h}_y = h_{ij}(x,y)dx^i \otimes dx^j$ on T_xM, which are defined by $h_{ij} := FF_{y^iy^j}$. It is clear that if F satisfies (5.6), then it satisfies (5.7). But the converse might be false although no counter-example has been found.

Example 5.1.3   ([89]) Let $\Theta = \Theta(x,y)$ denote the Funk metric on a strongly convex domain $\mathcal{U} \subset \text{R}^n$ (see Example 3.4.3). Let $\langle \, , \, \rangle$ denote the inner product on R^n and $a \in \text{R}^n$ be a constant vector. Let

$$F := \Theta(x,y) + \frac{\langle a, y \rangle}{1 + \langle a, x \rangle}, \qquad y \in T_x\mathcal{U} \cong \text{R}^n.$$

Since $F(0,y) = \Theta(0,y) + \langle a, y \rangle$ and $\Theta(0,y)$ is a Minkowski norm, by continuity, one can see that F is a Finsler metric in a neighborhood of the origin in $\mathbb{R}^n$ for sufficiently small vector a. According to Example 3.4.6, the spray coefficients of F are given by $G^i = Py^i$, where

$$P := \frac{1}{2}\left\{\Theta(x, y) - \frac{\langle a, y\rangle}{1 + \langle a, x\rangle}\right\}.$$

Thus F is projectively flat. A direct computation using the homogeneity of P gives

$$\frac{\partial G^m}{\partial y^m} = (n + 1)P.$$

Observe that

$$\mathbf{S} = (n + 1)P - y^m \frac{\partial}{\partial x^m}\left(\ln \sigma_F\right)$$

$$= \frac{n + 1}{2}F - (n + 1)\frac{\langle a, y\rangle}{1 + \langle a, x\rangle} - y^m \frac{\partial}{\partial x^m}\left(\ln \sigma_F\right)$$

$$= (n + 1)\left\{\frac{1}{2}F + df\right\},$$

where

$$f := -\ln\left[(1 + \langle a, x\rangle)\sigma_F(x)^{\frac{1}{n+1}}\right]. \tag{5.8}$$

Thus the S-curvature is almost constant. It follows from the above formula for $\mathbf{S}$ that

$$\mathbf{E} = \frac{1}{4}(n + 1)F^{-1}\mathbf{h}.$$

Thus the E-curvature is constant.

5.2   Randers Metrics of Isotropic S-Curvature

First of all, let us take a look at the following example.

Example 5.2.1   Let

$$F := \frac{\sqrt{|y|^2 - (|x|^2|y|^2 - \langle x, y\rangle^2)} + \langle x, y\rangle}{1 - |x|^2} + \frac{\langle a, y\rangle}{1 + \langle a, x\rangle}, \tag{5.9}$$

where $a \in \mathbb{R}^n$ is an arbitrary constant vector with $|a| < 1$. F is a Randers metric defined on the whole unit ball $\mathbb{B}^n(1) \subset \mathbb{R}^n$. By Example 5.1.3,

$G^i = P(x, y)y^i$ where $P = P(x, y)$ is given by

$$P = \frac{1}{2}\left\{\frac{\sqrt{|y|^2 - (|x|^2|y|^2 - \langle x, y\rangle^2)} + \langle x, y\rangle}{1 - |x|^2} - \frac{\langle a, y\rangle}{1 + \langle a, x\rangle}\right\}.$$

According to Example 1.3.1,

$$\sigma_F(x) = \frac{(1 - |a|^2)^{\frac{n+1}{2}}}{(1 + \langle a, x\rangle)^{n+1}}.$$

Plugging it into (5.8) yields that $h(x) = -\ln\sqrt{1 - |a|^2}$. Hence

$$\mathbf{S} = \frac{n+1}{2}F.$$

Thus F has constant S-curvature.

It is a natural problem to study and characterize Randers metrics with isotropic (or constant) S-curvature. Let $F = \alpha + \beta$ be a Randers metric on an n-dimensional manifold M, where $\alpha = \sqrt{a_{ij}(x)y^iy^j}$ and $\beta = b_i(x)y^i$. Let

$$\rho := \ln\sqrt{1 - \|\beta_x\|_\alpha^2}.$$

By (1.28), the volume forms dV_F and dV_α are related by

$$dV_F = e^{(n+1)\rho(x)}dV_\alpha.$$

We continue to use the same notations as in Lemma 3.1.2. The spray coefficients $G^i = G^i(x, y)$ of F and the spray coefficients $G^i_\alpha = G^i_\alpha(x, y)$ of α are related by (3.6) and (3.7),

$$G^i = G^i_\alpha + Py^i + Q^i,$$

where

$$P := \frac{e_{00}}{2F} - s_0, \qquad Q^i = \alpha s^i{}_0,$$

where $e_{00} := e_{ij}y^iy^j$, $s_0 := s_iy^i$ and $s^i{}_0 := s^i{}_jy^j$ are defined in (3.3) and (3.4). Since $s_{ij} + s_{ji} = 0$, we have that $s_{00} := s_{ij}y^iy^j = 0$ and

$s^i{}_i = a^{ij}s_{ij} = 0$. By the homogeneity of P and the anti-symmetry of s_{ij}, we have

$$\frac{\partial(Py^m)}{\partial y^m} = \frac{\partial P}{\partial y^m}y^m + nP = (n+1)P,$$

$$\frac{\partial Q^m}{\partial y^m} = \alpha^{-1}s_{00} + \alpha s^m{}_m = 0.$$

Since α is Riemannian, the following holds,

$$\frac{\partial G^m_\alpha}{\partial y^m} = \bar{\Gamma}^m_{im}y^i = y^m\frac{\partial}{\partial x^m}\left(\sqrt{\det(a_{ij})}\right) = y^m\frac{\partial}{\partial x^m}\left(\ln\sigma_\alpha\right),$$

where $\bar{\Gamma}^i_{jk}$ denote the Christoffel symbols of α. By the above identities, we obtain

$$\begin{aligned}
\mathbf{S} &= \frac{\partial G^m}{\partial y^m} - y^m\frac{\partial}{\partial x^m}\left(\ln\sigma_F\right) \\
&= \frac{\partial G^m_\alpha}{\partial y^m} + \frac{\partial(Py^m)}{\partial y^m} + \frac{\partial Q^m}{\partial y^m} - (n+1)y^m\frac{\partial\rho}{\partial x^m} - y^m\frac{\partial}{\partial x^m}\left(\ln\sigma_\alpha\right) \\
&= (n+1)\left\{P - \rho_0\right\} \\
&= (n+1)\left\{\frac{e_{00}}{2F} - (s_0 + \rho_0)\right\},
\end{aligned} \qquad (5.10)$$

where $\rho_0 := \rho_{x^i}(x)y^i$.

Lemma 5.2.2   ([25]) *For a Randers metric $F = \alpha+\beta$ on an n-dimensional manifold M, the following are equivalent*

(a) $\mathbf{S} = (n+1)cF$,
(b) $\mathbf{E} = \frac{1}{2}(n+1)cF^{-1}\mathbf{h}$,
(c) $e_{00} = 2c(\alpha^2 - \beta^2)$,

where $c = c(x)$ is a scalar function on M.

Proof. From the definitions of $\mathbf{S}$ and $\mathbf{E}$, it is obvious that (a) $\Rightarrow$ (b).

(b) $\Rightarrow$ (c): The condition (b) implies that

$$\mathbf{S} = (n+1)\left\{c(x)F + \eta\right\},$$

where $\eta = \eta_i(x)y^i$ is a 1-form on M. By (5.10), (b) is equivalent to the following

$$e_{00} = 2cF^2 + 2\theta F,$$

where $\theta := s_0 + \rho_0 + \eta$. This implies that

$$e_{00} = 2c(\alpha^2 + \beta^2) + 2\theta\beta, \qquad 0 = 4c\beta + 2\theta.$$

Solving for θ from the above equation on the right, then substituting it into the equation on the left, we obtain (c).

(c) $\Rightarrow$ (a): Substituting $e_{00} = 2c(\alpha^2 - \beta^2)$ into (5.10) yields

$$\mathbf{S} = (n+1)\Big\{ c(\alpha - \beta) - (s_0 + \rho_0) \Big\}. \tag{5.11}$$

On the other hand, contracting $e_{ij} = 2c(a_{ij} - b_i b_j)$ with b^j gives

$$s_i + \rho_i + 2cb_i = 0.$$

Thus $s_0 + \rho_0 = -2c\beta$. Substituting it into (5.11) yields (a).   Q.E.D.

Example 5.2.3 Consider the Randers metric $F = \alpha + \beta$ on $\mathbf{R}^n$, where α and β are defined by

$$\alpha := \frac{\sqrt{(1-\varepsilon^2)\langle x, y\rangle^2 + \varepsilon|y|^2(1+\varepsilon|x|^2)}}{1 + \varepsilon|x|^2}$$

$$\beta := \frac{\sqrt{1-\varepsilon^2}\,\langle x, y\rangle}{1 + \varepsilon|x|^2},$$

where ε is an arbitrary constant with $|\varepsilon| < 1$. Note that β is closed. Thus $s_{ij} = 0$ and $s_i = 0$. By computing $b_{i;j}$, one obtains

$$e_{ij} = \frac{\varepsilon\sqrt{1-\varepsilon^2}}{(1+\varepsilon|x|^2)(\varepsilon+|x|^2)}\delta_{ij}.$$

On the other hand,

$$a_{ij} - b_i b_j := \frac{\varepsilon}{1+\varepsilon|x|^2}\delta_{ij}.$$

Thus $e_{ij} = 2c(a_{ij} - b_i b_j)$ with

$$c := \frac{\sqrt{1-\varepsilon^2}}{2(\varepsilon+|x|^2)}.$$

By Lemma 5.2.2, F has isotropic S-curvature and E-curvature, i.e.,

$$\mathbf{S} = (n+1)cF, \qquad \mathbf{E} = \frac{1}{2}cF^{-1}\mathbf{h}.$$

Since α is not projectively flat and β is closed when $\varepsilon \neq 0$, by Proposition 3.4.8, one can see that F is not projectively flat.

It is known that a Finsler metric $F = F(x, y)$ on a manifold M is a Randers metric if and only if it is a solution of the following equation:

$$h\left(x, \frac{y}{F} - V_x\right) = 1,$$

where $h = \sqrt{h_{ij}(x)y^iy^j}$ is a Riemannian metric and $V = V^i(x)\frac{\partial}{\partial x^i}$ is a vector field with $h(x, -V_x) = \sqrt{h_{ij}(x)V^i(x)V^j(x)} < 1$. F is given by (3.8),

$$F = \frac{\sqrt{\lambda h^2 + V_0}}{\lambda} - \frac{V_0}{\lambda}, \tag{5.12}$$

where $V_0 := V_iy^i$ and $\lambda = 1 - \|V\|_h^2$.

Let $\mathcal{R}_{ij} := \frac{1}{2}(V_{i|j} + V_{j|i})$, $\mathcal{R}_j := V^i\mathcal{R}_{ij}$, $\mathcal{R} := V^i\mathcal{R}_{ij}V^j$ be defined as in Lemma 3.1.3. By a direct computation, we obtain from (3.11) that

$$\frac{\partial G^m}{\partial y^m} = \frac{\partial G_h^m}{\partial y^m} + \frac{n+1}{2F}\left\{2F\mathcal{R}_0 - \mathcal{R}_{00} - F^2\mathcal{R}\right\}. \tag{5.13}$$

Let $dV_F = \sigma_F dx^1 \cdots dx^n$ and $dV_h = \sigma_h$ denote the volume form of F and h respectively. By Lemma 1.4.2, $dV_F = dV_h$, i.e., $\sigma_F = \sigma_h$. Since h is a Riemannian metric, we have

$$\frac{\partial G_h^m}{\partial y^m} = y^m \frac{\partial}{\partial x^m}\left(\ln \sigma_h\right). \tag{5.14}$$

Then it follows from (5.4), (5.13) and (5.14) that

$$\mathbf{S} = \frac{n+1}{2F}\left\{2F\mathcal{R}_0 - \mathcal{R}_{00} - F^2\mathcal{R}\right\}. \tag{5.15}$$

Let

$$\xi^i := y^i - F(x, y)V^i.$$

Since $\|V\|_h < 1$, the vector $\xi := \xi^i\frac{\partial}{\partial x^i}|_x \in T_xM$ can be arbitrary. Moreover, by (3.10), it is easy to verify that

$$h_{ij}\xi^i\xi^j = h^2 - 2FV_0 + (1 - \lambda)F^2 = F^2. \tag{5.16}$$

By (5.15) , we obtain the following

Lemma 5.2.4 *For an n-dimensional Randers metric F expressed by (5.12), the S-curvature is given by*

$$\frac{\mathbf{S}}{F} = -\frac{n+1}{2}\frac{\mathcal{R}_{ij}\xi^i\xi^j}{h_{ij}\xi^i\xi^j}. \tag{5.17}$$

By (5.17), we immdiately obtain the following

Proposition 5.2.5 *Let $F = \alpha + \beta$ be an n-dimensional Randers metric expressed in terms of a Riemannian metric $h = \sqrt{h_{ij}(x)y^iy^j}$ and a vector field $V = V^i(x)\frac{\partial}{\partial x^i}$ by (5.12). Then F has isotropic S-curvature, $\mathbf{S} = (n+1)c(x)F$ if and only if V satisfies*

$$\mathcal{R}_{ij} = -2c(x)h_{ij}. \tag{5.18}$$

By Lemma 5.2.2, the condition $\mathbf{S} = (n+1)c(x)F$ is equivalent to the following equation

$$e_{00} = 2c(x)(\alpha^2 - \beta^2). \tag{5.19}$$

It is implicitly shown in [81], [9] and [7] that (5.19) is equivalent to (5.18). But the arguments are embedded in the proof of the theorem on Randers metrics of constant flag curvature. Under the constant flag curvature condition, (5.18) and (5.19) hold for a constant c. A direct proof of Proposition 5.2.5 was first given by then an undergraduate student at Beijing University, H. Xing, shortly after he received the paper [7]. Xing's argument does not use (5.17).

In what follows, we are going to give explicit formulas for $V = V^i\frac{\partial}{\partial x^i}$ and $c = c(x)$ satisfying (5.18) when the underlying Riemannian metric h is of constant sectional curvature [94]. First we prove the following

Lemma 5.2.6 *Let $h = \sqrt{h_{ij}y^iy^j}$ be a Riemannian metric on an n-dimensional manifold M. Let $V = V^i\frac{\partial}{\partial x^i}$ be a vector field on M satisfying (5.18) for some scalar function $c = c(x)$ on M. Then*

$$V_{k|i|j} = 2\Big\{ c_{|k}h_{ij} - c_{|i}h_{jk} - c_{|j}h_{ki} \Big\} - V_m\bar{R}_j{}^m{}_{ki}, \tag{5.20}$$

where $c_{|i} = c_{x^i}$ are the usual partial derivatives of c in local coordinate systems and $\bar{R}_j{}^m{}_{ki}$ denote the coefficients of the Riemann curvature tensor of h.

Proof: First, differentiating (5.18) and exchanging the indices, we obtain

$$V_{i|j|k} + V_{j|i|k} = -4c_{|k}h_{ij}, \tag{5.21}$$

$$V_{j|k|i} + V_{k|j|i} = -4c_{|i}h_{jk}, \tag{5.22}$$

$$V_{k|i|j} + V_{i|k|j} = -4c_{|j}h_{ki}. \tag{5.23}$$

Adding (5.22) and (5.23) together, then the sum being subtracted by (5.21), we obtain

$$(V_{i|k|j} - V_{i|j|k}) + (V_{j|k|i} - V_{j|i|k}) + (V_{k|j|i} - V_{k|i|j}) + 2V_{k|i|j}$$

$$= 4c_{|k}h_{ij} - 4c_{|i}h_{jk} - 4c_{|j}h_{ki}.$$

Using the Ricci identity, $V_{k|i|j} - V_{k|j|i} = V_m \bar{R}_k{}^m{}_{ij}$, we obtain

$$V_m \bar{R}_i{}^m{}_{kj} + V_m \bar{R}_j{}^m{}_{ki} + V_m \bar{R}_k{}^m{}_{ji} + 2V_{k|i|j} = 4c_{|k}h_{ij} - 4c_{|i}h_{jk} - 4c_{|j}h_{ki}.$$

By applying the Bianchi identity, $\bar{R}_i{}^m{}_{kj} + \bar{R}_k{}^m{}_{ji} + \bar{R}_j{}^m{}_{ik} = 0$, we obtain (5.20).
$$\text{Q.E.D.}$$

By Lemma 5.2.6, we obtain another identity on c and V.

Lemma 5.2.7   *Let (M, h) be an n-dimensional Riemannian manifold. Let $V = V^i \frac{\partial}{\partial x^i}$ be a vector field on M and $c = c(x)$ be a scalar function on M satisfying (5.18) for some scalar function $c = c(x)$. Then c and V satisfy*

$$2\Big\{c_{|i|l}h_{jk} + c_{|j|k}h_{li}\Big\} - 2\Big\{c_{|j|l}h_{ki} + c_{|i|k}h_{jl}\Big\}$$

$$= 4c\bar{R}_{ijkl} + V_m\Big\{\bar{R}_k{}^m{}_{ij|l} - \bar{R}_l{}^m{}_{ij|k}\Big\}$$

$$+ V_{m|j}\bar{R}_i{}^m{}_{kl} - V_{m|i}\bar{R}_j{}^m{}_{kl} + V_{m|l}\bar{R}_k{}^m{}_{ij} - V_{m|k}\bar{R}_l{}^m{}_{ij}. \tag{5.24}$$

Proof: By (5.20),

$$V_{i|j|k} = 2\Big\{c_{|i}h_{jk} - c_{|j}h_{ki} - c_{|k}h_{ij}\Big\} - V_m\bar{R}_k{}^m{}_{ij}. \tag{5.25}$$

Differentiating (5.25) yields

$$V_{i|j|k|l} = 2\Big\{c_{|i|l}h_{jk} - c_{|j|l}h_{ki} - c_{|k|l}h_{ij}\Big\} - V_{m|l}\bar{R}_k{}^m{}_{ij} - V_m\bar{R}_k{}^m{}_{ij|l}. \tag{5.26}$$

Exchanging the indices k and l yields

$$V_{i|j|l|k} = 2\Big\{c_{|i|k}h_{jl} - c_{|j|k}h_{li} - c_{|l|k}h_{ij}\Big\} - V_{m|k}\bar{R}_l{}^m{}_{ij} - V_m\bar{R}_l{}^m{}_{ij|k}. \tag{5.27}$$

Note that $c = c(x)$ is a scalar function, thus $c_{|k|l} = c_{|l|k}$. It follows from (5.26) and (5.27) that

$$V_{i|j|k|l} - V_{i|j|l|k} = 2\Big\{c_{|i|l}h_{jk} + c_{|j|k}h_{li}\Big\} - 2\Big\{c_{|j|l}h_{ki} + c_{|i|k}h_{jl}\Big\}$$
$$+ V_{m|k}\bar{R}_l{}^m{}_{ij} - V_{m|l}\bar{R}_k{}^m{}_{ij} + V_m\Big\{\bar{R}_l{}^m{}_{ij|k} - \bar{R}_k{}^m{}_{ij|l}\Big\}.$$

Applying the Ricci identity $V_{i|j|k|l} - V_{i|j|l|k} = V_{m|j}\bar{R}_i{}^m{}_{kl} + V_{i|m}\bar{R}_j{}^m{}_{kl}$ and the identity (5.18) to the above identity, we obtain

$$2\Big\{c_{|i|l}h_{jk} + c_{|j|k}h_{li}\Big\} - 2\Big\{c_{|j|l}h_{ki} + c_{|i|k}h_{jl}\Big\}$$
$$= 4c\bar{R}_{ijkl} + V_m\Big\{\bar{R}_k{}^m{}_{ij|l} - \bar{R}_l{}^m{}_{ij|k}\Big\}$$
$$+ V_{m|j}\bar{R}_i{}^m{}_{kl} - V_{m|i}\bar{R}_j{}^m{}_{kl} + V_{m|l}\bar{R}_k{}^m{}_{ij} - V_{m|k}\bar{R}_l{}^m{}_{ij}.$$

This gives (5.24). Q.E.D.

Lemma 5.2.8 *Let $h = \sqrt{h_{ij}y^iy^j}$ be a Riemannian metric of constant curvature $\mathbf{K}_h = \mu$ and $V = V^i\frac{\partial}{\partial x^i}$ be a vector field on an n-dimensional manifold M satisfying (5.18) for some scalar function $c = c(x)$. Then c satisfies the following equations*

$$c_{|i|j} + \mu ch_{ij} = 0, \qquad (n > 2) \tag{5.28}$$
$$\Delta c + 2\mu c = 0, \qquad (n = 2) \tag{5.29}$$

where $c_{|i|j}$ denote the covariant derivatives of c with respect to h and $\Delta c := h^{ij}c_{|i|j}$ denotes the Laplacian of c with respect to h.

Proof. By assumption, h has constant curvature,

$$\bar{R}_k{}^m{}_{ij} = \mu\Big\{\delta_i^m h_{jk} - \delta_j^m h_{ik}\Big\}.$$

By (5.18), we obtain from (5.24) that

$$\Big\{c_{|i|l}h_{jk} + c_{|j|k}h_{li}\Big\} - \Big\{c_{|j|l}h_{ki} + c_{|i|k}h_{jl}\Big\} = 2\mu c\Big\{h_{jl}h_{ik} - h_{jk}h_{il}\Big\}. \tag{5.30}$$

For the sake of simplicity, we may choose an orthonormal basis at a point so that $h_{ij} = \delta_{ij}$. In (5.24), letting $k = j$ and $l = i$ ($i \neq j$) yields

$$c_{|i|i} + c_{|j|j} + 2\mu c = 0 \qquad (i \neq j). \tag{5.31}$$

When $n \geq 3$, it follows from (5.31) that

$$c_{|i|i} + \mu c = 0. \tag{5.32}$$

For any i, l, there is $m \neq i, l$. In (5.24), letting $j = k = m$, one obtains

$$c_{|i|l} + c_{|m|m}\delta_{il} + 2\mu c\delta_{il} = 0. \tag{5.33}$$

By (5.32), $c_{|m|m} = -\mu c$. Substituting it into (5.33) yields (5.28). In dimension two, (5.29) follows from (5.31) directly.   Q.E.D.

Next we are going to find an explicit formula for the vector field satisfying (5.18) when the Riemannian metric h is of constant curvature $\mathbf{K}_h = \mu$. It is known that h is locally isometric to the following metric $h_\mu = \sqrt{h_{ij}y^i y^j}$ on the ball $\mathrm{B}^n(r_\mu) \subset \mathrm{R}^n$, where $r_\mu := +\infty$ if $\mu \geq 0$ and $r_\mu := \pi/\sqrt{-\mu}$,

$$h_\mu = \frac{\sqrt{|y|^2 + \mu(|x|^2|y|^2 - \langle x, y\rangle^2)}}{1 + \mu|x|^2}, \qquad y \in T_x\mathrm{B}^n(r_\mu) \cong \mathrm{R}^n. \tag{5.34}$$

The metric coefficients h_{ij} are given by

$$h_{ij} := \frac{\delta_{ij}}{1 + \mu|x|^2} - \frac{\mu x^i x^j}{(1 + \mu|x|^2)^2}.$$

The inverse matrix of (h_{ij}) is given by

$$h^{ij} = (1 + \mu|x|^2)\Big\{\delta^{ij} + \mu x^i x^j\Big\}.$$

Then the Christoffel symbols of γ^i_{jk} of h_μ are given

$$\gamma^i_{jk} = -\mu\frac{x^j\delta^i_k + x^k\delta^i_j}{1 + \mu|x|^2}.$$

Thus for a tensor $T = T_i dx^i$, the covariant derivative $DT = T_{i|j}dx^i \otimes dx^j$ is given by

$$T_{i|j} = \frac{\partial T_i}{\partial x^j} + \mu\frac{x^i T_j + x^j T_i}{1 + \mu|x|^2}. \tag{5.35}$$

Lemma 5.2.9 *If a scalar function $c = c(x)$ satisfies (5.28) on the Euclidean ball $(\mathrm{B}^n(r_\mu), h_\mu)$, then*

$$c = \frac{\lambda + \langle a, x\rangle}{\sqrt{1 + \mu|x|^2}}, \tag{5.36}$$

where λ is a constant and $a \in \mathbb{R}^n$ is a constant vector.

Proof: By (5.35), we obtain

$$c_{|i|j} = c_{x^i x^j} + \mu \frac{x^i c_{x^j} + x^j c_{x^i}}{1 + \mu |x|^2},$$

where $c_{x^i} = \frac{\partial c}{\partial x^i}$ and $c_{x^i x^j} = \frac{\partial^2 c}{\partial x^i \partial x^j}$ denote the partial derivatives of c. Let $f := \sqrt{1 + \mu |x|^2}\, c$. We have

$$f_{x^i x^j} = \sqrt{1 + \mu |x|^2} \Big\{ c_{|i|j} + \mu c h_{ij} \Big\} = 0.$$

Thus

$$f = \lambda + \langle a, x \rangle,$$

where λ is a constant and $a \in \mathbb{R}^n$ is a constant vector. We obtain the explicit formula (5.36) for c.   Q.E.D.

For the above scalar function c in (5.36), we can solve (5.18) for V.

Proposition 5.2.10 ([94]) *Let $h = h_\mu$ be defined in (7.38) and V be a vector field on the Euclidean ball $(\mathrm{B}^n(r_\mu), h_\mu)$. If $n \geq 3$, $V = V^i \frac{\partial}{\partial x^i}$ satisfies (5.18) for some scalar function $c = c(x)$, then c is given by (5.36) and V is given by*

$$V = -2 \Big\{ \Big(\lambda \sqrt{1 + \mu |x|^2} + \langle a, x \rangle \Big) x - \frac{|x|^2 a}{\sqrt{1 + \mu |x|^2} + 1} \Big\}$$
$$+ xQ + b + \mu \langle b, x \rangle x, \tag{5.37}$$

where $Q = (q_j{}^i)$ is an antisymmetric matrix and $b = (b^i) \in \mathbb{R}^n$ is a constant vector. Conversely, for any dimension $n \geq 2$, the vector field $V = (V^i)$ in (7.40) satisfies (5.18) for the scalar function $c = c(x)$ in (5.36).

Proof: It suffices to solve (5.18) for the scalar function in (5.36). We divide the argument into two cases.

Case 1: $\mu = 0$. Let

$$P_i := V_i - |x|^2 a^i + 2(\lambda + \langle a, x \rangle) x^i.$$

Then (5.18) is equivalent to

$$\frac{\partial P_i}{\partial x^j} + \frac{\partial P_j}{\partial x^i} = 0.$$

By an elementary argument (see [7]), we get

$$P_i = x^j q_j{}^i + b^i,$$

where $Q = (q_j{}^i)$ is an antisymmetric matrix and $b = (b^i) \in \mathbf{R}^n$ is a constant vector. We obtain

$$V^i = V_i = -2\left\{(\lambda + \langle a, x \rangle)x^i - \frac{|x|^2}{2}a^i\right\} + x^j q_j{}^i + b^i.$$

Case 2: $\mu \neq 0$. Let

$$P_i := V_i - \frac{2}{\mu}c_{;i}.$$

Then P_i satisfy

$$P_{i|j} + P_{j|i} = 0. \tag{5.38}$$

Using (5.35), we can rewrite (5.38) as follows:

$$\frac{\partial P_i}{\partial x^j} + \frac{\partial P_j}{\partial x^i} + 2\mu \frac{x^i P_j + x^j P_i}{1 + \mu|x|^2} = 0.$$

Let $H_i := (1 + \mu|x|^2)P_i$. We obtain

$$\frac{\partial H_i}{\partial x^j} + \frac{\partial H_j}{\partial x^i} = (1 + \mu|x|^2)\left\{\frac{\partial P_i}{\partial x^j} + \frac{\partial P_j}{\partial x^i} + 2\mu \frac{x^j P_i + x^i P_j}{1 + \mu|x|^2}\right\} = 0.$$

By an argument similar to that for P_i in the case when $\mu = 0$, we obtain

$$H_i = x^j q_j{}^i + v^i,$$

where $Q = (q_j{}^i)$ is an antisymmetric matrix and $v = (v^i) \in \mathbf{R}^n$ is a constant vector. Thus

$$P_i = (1 + \mu|x|^2)^{-1}\left\{x^j q_j{}^i + v^i\right\}.$$

A direct computation yields

$$c_{|i} = \frac{a^i}{\sqrt{1 + \mu|x|^2}} - \frac{\mu(\lambda + \langle a, x \rangle)x^i}{(1 + \mu|x|^2)^{3/2}}. \tag{5.39}$$

We obtain

$$V_i = P_i + \frac{2}{\mu}c_{|i}$$

$$= (1 + \mu|x|^2)^{-1}\left\{x^j q_j{}^i + v^i\right\} + \frac{2a^i}{\mu\sqrt{1 + \mu|x|^2}} - \frac{2(\lambda + \langle a, x\rangle)x^i}{(1 + \mu|x|^2)^{3/2}}.$$

Finally, we completely determine $V^i = h^{ij}V_j$.

$$V^i = 2\sqrt{1 + \mu|x|^2}\left\{\mu^{-1}a^i - \lambda x^i\right\} + x^j q_j{}^i + v^i + \mu\langle v, x\rangle x^i.$$

We express

$$\mu^{-1}\sqrt{1 + \mu|x|^2} = \mu^{-1}\left\{\sqrt{1 + \mu|x|^2} - 1\right\} + \mu^{-1}$$

$$= \frac{|x|^2}{\sqrt{1 + \mu|x|^2} + 1} + \mu^{-1}.$$

Let

$$b^i := v^i + 2\mu^{-1}a^i.$$

We obtain

$$V^i = -2\left(\lambda\sqrt{1 + \mu|x|^2} + \langle a, x\rangle\right)x^i + \frac{2|x|^2 a^i}{\sqrt{1 + \mu|x|^2} + 1}$$

$$+ x^j q^i{}_j + b^i + \mu\langle b, x\rangle x^i.$$

Q.E.D.

Assume that a vector field V satisfies (5.18) on the Euclidean ball $(\mathrm{B}^n(r_\mu), h_\mu)$ for some constant c. By Lemma 5.2.8, we see that $c = 0$ if $\mu \neq 0$. Then (5.28) always holds regardless of dimension. By Lemma 5.2.9, c can be expressed by

$$c = \frac{\lambda + \langle a, x\rangle}{\sqrt{1 + \mu|x|^2}},$$

where $\lambda = 0$ and $a = 0$ if $\mu \neq 0$, and $\lambda = c$ and $a = 0$ if $\mu = 0$. By the argument of Proposition 5.2.10, we obtain the following

Corollary 5.2.11 *Let $h = h_\mu$ be defined in (5.34) and V be a vector field on the Euclidean ball $(\mathrm{B}^n(r_\mu), h_\mu)$. If $V = V^i\frac{\partial}{\partial x^i}$ satisfies (5.18) for some*

constant c, then $c = 0$ if $\mu \neq 0$, and V is given by

$$V = \begin{cases} -2cx + xQ + b & \text{if } \mu = 0 \\ xQ + b + \mu\langle a, x\rangle x & \text{if } \mu \neq 0 \end{cases}, \tag{5.40}$$

where Q is an anti-symmetric matrix and $a, b \in \mathbf{R}^n$ are constant vectors. Conversely, if V is a vector field satisfying (5.40), then V satisfies (5.18) with $c(x) = c$ $(= 0$ if $\mu \neq 0)$.

5.3   An Equation on the S-Curvature

The main purpose of this section is to establish an equation on the S-curvature. This equation together with (2.70) gives a relation between the S-curvature and the Riemann tensor.

Lemma 5.3.1   ([24], [73])

$$\mathbf{S}_{\cdot k|m}y^m - \mathbf{S}_{|k} = I_{k|p|q}y^p y^q + I_m R^m{}_k, \tag{5.41}$$

$$\mathbf{S}_{\cdot k|m}y^m - \mathbf{S}_{|k} = -\frac{1}{3}\Big\{ 2R^m{}_{k \cdot m} + R^m{}_{m \cdot k} \Big\}. \tag{5.42}$$

Proof. Equation (5.42) follows from (2.70) and (5.41). Thus it suffices to prove (5.41). Take a natural local frame $\{\mathbf{e}_i = \partial_i\}$ for $\pi^* TM$. Let $\{\omega^i, \omega^{n+i}\} = \{dx^i, \delta y^i\}$ be the corresponding local coframe for $T^*(TM_o)$. Write

$$d\tau = \tau_{|i}\omega^i + \tau_{\cdot i}\omega^{n+i}. \tag{5.43}$$

It follows from (5.2) that

$$\tau_{\cdot i} = I_i. \tag{5.44}$$

Differentiating (5.43) and using (5.44), we obtain

$$0 = d^2\tau = \Big\{ \tau_{|k|l}\omega^l + \tau_{|k \cdot l}\omega^{n+l} \Big\} \wedge \omega^k$$
$$+ \Big\{ I_{k|l}\omega^l + I_{k \cdot l}\omega^{n+l} \Big\} \wedge \omega^k + I_m\Omega^{n+m}.$$

This yields the following Ricci identities

$$\tau_{|k|l} = \tau_{|l|k} + I_m R^m{}_{kl}, \tag{5.45}$$

$$\tau_{|k \cdot l} = I_{l|k} - I_m L^m{}_{kl}. \tag{5.46}$$

From the definition of the S-curvature, we have

$$\mathbf{S} = \tau_{|m} y^m. \tag{5.47}$$

Contracting (5.46) with y^k and using (5.47), we obtain

$$
\begin{aligned}
\mathbf{S}_{\cdot k} = (\tau_{|m} y^m)_{\cdot k} &= \tau_{|m \cdot k} y^m + \tau_{|k} \\
&= I_{k|m} y^m - I_l L^l_{mk} y^m + \tau_{|k} \\
&= I_{k|m} y^m + \tau_{|k} = J_k + \tau_{|k}.
\end{aligned}
\tag{5.48}
$$

It follows from (5.48) that

$$\mathbf{S}_{\cdot k|l} = \tau_{|k|l} + J_{k|l}. \tag{5.49}$$

By (5.45) and (5.49), we obtain

$$
\begin{aligned}
\mathbf{S}_{\cdot k|m} y^m - \mathbf{S}_{|k} &= \left\{ \mathbf{S}_{\cdot k|m} - \mathbf{S}_{\cdot m|k} \right\} y^m \\
&= \left\{ \tau_{|k|m} - \tau_{|m|k} \right\} y^m + \left\{ J_{k|m} - J_{m|k} \right\} y^m \\
&= I_m R^m_{\ k} + J_{k|m} y^m.
\end{aligned}
$$

Since $J_{k|m} y^m = I_{k|p|q} y^p y^q$, we obtain (5.41). Q.E.D.

Equations (5.41) and (5.42) reveal some important relationships among the S-curvature and the Riemann tensor. We shall use them to establish some global rigidity theorems for Finsler metrics.

Chapter 6

Riemann Curvature

In the previous chapter, we discussed some non-Riemannian quantities. These quantities all vanish for Berwald metrics. In particular, they vanish for Riemannian metrics. In Riemannian geometry, the Riemann tensor is the most important quantity that measures the "curvature" of the space at a point. The Riemann tensor is first extended by L. Berwald to Finsler metrics [14], [15]. Our goal is to understand the geometric meaning of this important quantity.

6.1 Riemann Curvature

Let (M, F) be an n-dimensional Finsler manifold. Let $\{\mathbf{b}_i\}$ be a local frame for TM and $\{\theta^i\}$ be the local coframe for T^*M. Then $\{\mathbf{e}_i := (x, y, \mathbf{b}_i)\}$ is a local frame for π^*TM and $\{\omega^i := \pi^*\theta^i\}$ is the dual local coframe for π^*T^*M. The Riemann tensor $R = R^i{}_k \mathbf{e}_i \otimes \omega^k$ is defined in (2.43) using the Chern connection. It can be viewed as a family of linear maps

$$\mathbf{R} = \Big\{ \mathbf{R}_y \mid y \in T_x M \setminus \{0\}, \ x \in M \Big\},$$

where $\mathbf{R}_y = R^i{}_k \mathbf{b}_i \otimes \theta^k : T_x M \to T_x M$ is defined by

$$\mathbf{R}_y(v) := R^i{}_k(x, y) v^k \, \mathbf{b}_i, \quad v = v^i \mathbf{b}_i \in T_x M.$$

$\mathbf{R}$ is called the *Riemann curvature*. We have

$$\mathbf{R}_y(y) = 0, \quad \mathbf{R}_{\lambda y} = \lambda^2 \mathbf{R}_y, \quad \lambda > 0. \tag{6.1}$$

107

By (2.44), $\mathbf{R}_y$ is self-adjoint with respect to $\mathbf{g}_y$,

$$\mathbf{g}_y\Big(\mathbf{R}_y(u),\ v\Big) = \mathbf{g}_y\Big(u,\ \mathbf{R}_y(v)\Big), \qquad u, v \in T_x M. \tag{6.2}$$

(6.1) and (6.2) are the basic equations of the Riemann curvature.

For a tangent plane $P \subset T_x M$ containing y, let

$$\mathbf{K}(P, y) := \frac{\mathbf{g}_y\Big(\mathbf{R}_y(u),\ u\Big)}{\mathbf{g}_y(y, y)\mathbf{g}_y(u, u) - [\mathbf{g}_y(y, u)]^2}, \tag{6.3}$$

where $u \in P$ such that $P = \mathrm{span}\{y, u\}$.

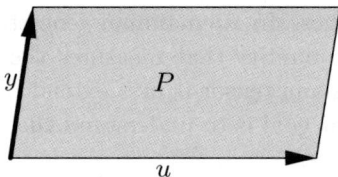

Figure 6.1

By (6.1) and (6.2), one can easily verify that $\mathbf{K}(P, y)$ is independent of the choice of a particular vector $u \in P$ such that $P = \mathrm{span}\{y, u\}$. $\mathbf{K} = \mathbf{K}(P, y)$ is called the *flag curvature*. In dimension two, $P = T_x M$ is the tangent plane. Thus the flag curvature $\mathbf{K} = \mathbf{K}(x, y)$ is a scalar function on $TM \backslash \{0\}$, which is called the *Gauss curvature*.

Let

$$\mathbf{Ric} := \sum_{i=1}^{n} g^{ij}\mathbf{g}_y\Big(\mathbf{R}_y(\mathbf{b}_i), \mathbf{b}_j\Big),$$

where $\{\mathbf{b}_i\}$ is a basis for $T_x M$, $g_{ij} := \mathbf{g}_y(\mathbf{b}_i, \mathbf{b}_j)$ and $(g^{ij}) = (g_{ij})^{-1}$. $\mathbf{Ric}$ is a well-defined scalar function on $TM \backslash \{0\}$. We call $\mathbf{Ric}$ the *Ricci curvature* (or *Ricci scalar*). In a local coordinate system,

$$\mathbf{Ric} = g^{ij} R_{ij} = R^m{}_m.$$

Proposition 6.1.1 *Let F be a Riemannian metric on a manifold M. For any tangent plane $P \subset T_x M$, the flag curvature $\mathbf{K}(P, y) = \mathbf{K}(P)$ is independent of $y \in P \setminus \{0\}$.*

Proof: Since F is Riemannian ($C_{ijk} = 0$), $g_{ij} = g_{ij}(x)$ and $R_j{}^i{}_{kl} = R_j{}^i{}_{kl}(x)$ are functions of $x \in M$. Thus $R_{jikl} := g_{im} R_j{}^m{}_{kl}$ are functions of $x \in M$. It follows from (2.34) and (2.64) that

$$R_{jikl} = -R_{ijkl} = R_{ijlk}.$$

This implies that

$$R_{ik}(x, y) u^i u^k = R_{jikl}(x) y^j y^l u^i u^k = R_{ijlk}(x) u^i u^k y^j y^l = R_{jl}(x, u) y^j y^l.$$

Thus for any tangent plane $P = \mathrm{span}\{y, u\} \subset T_x M$,

$$
\begin{aligned}
\mathbf{K}(P, y) &= \frac{R_{ik}(x, y) u^i u^k}{\{g_{jl}(x) g_{ik}(x) - g_{ij}(x) g_{kl}(x)\} y^j y^l u^i u^k} \\
&= \frac{R_{jl}(x, u) y^j y^l}{\{g_{jl}(x) g_{ik}(x) - g_{ij}(x) g_{kl}(x)\} y^j y^l u^i u^k} = \mathbf{K}(P, u).
\end{aligned}
$$

That is, the flag curvature $\mathbf{K}(P, y) = \mathbf{K}(P)$ is independent of $y \in P \setminus \{0\}$. Q.E.D.

For Riemannian metrics, the flag curvature $\mathbf{K} = \mathbf{K}(P)$ is called the *sectional curvature* of the section $P \subset T_x M$. In dimension two, the *Gauss curvature* $\mathbf{K} = \mathbf{K}(x)$ is a scalar function on M.

Definition 6.1.2 Let F be a Finsler metric on an n-dimensional manifold M. F is said to be *of scalar (flag) curvature* if $\mathbf{K}(P, y) = \mathbf{K}(x, y)$ is a scalar function on $TM \setminus \{0\}$. F is said to have isotropic flag curvature if $\mathbf{K}(P, y) = \mathbf{K}(x)$ is a scalar function on M. F is said to have *constant flag curvature* if $\mathbf{K}(P, y) = constant$. F is called an *Einstein metric* if there is a scalar function $\mathbf{K} = \mathbf{K}(x)$ on M such that

$$\mathbf{Ric} = (n - 1) \mathbf{K} F^2.$$

It follows from the above definition that a Finsler metric $F = F(x, y)$ on a manifold M is of scalar flag curvature with flag curvature $\mathbf{K} = \mathbf{K}(x, y)$

if and only if for any $y, u \in T_x M \setminus \{0\}$,

$$\mathbf{R}_y(u) = \mathbf{K} \left\{ \mathbf{g}_y(y, y)\, u - \mathbf{g}_y(y, u)\, y \right\}.$$

This is equivalent to the following equation:

$$R^i{}_k = \mathbf{K} F^2 h^i{}_k,$$

where $h^i{}_k := \delta^i_k - F^{-2} g_{kq} y^q y^i$. Compare (2.51).

There are plenty of Finsler metrics of scalar flag curvature. First, let us prove the following

Proposition 6.1.3 ([17]) *Any locally projectively flat Finsler metric is of scalar flag curvature.*

Proof. Let F be a locally projectively flat Finsler metric on a manifold M. By definition, at any point $x \in M$, there is a standard local coordinate system (x^i, y^i) in TM such that the spray coefficients are in the following form $G^i = P y^i$. Substituting them into (2.49) yields

$$
\begin{aligned}
R^i{}_k &= 2 \frac{\partial(P y^i)}{\partial x^k} - y^j \frac{\partial^2(P y^i)}{\partial x^j \partial y^k} + 2(P y^j) \frac{\partial^2(P y^i)}{\partial y^j \partial y^k} - \frac{\partial(P y^i)}{\partial y^j} \frac{\partial(P y^j)}{\partial y^k} \\
&= \left\{ P^2 - y^j P_{x^j} \right\} \delta^i_k + \left\{ 2 P_{x^k} - y^j P_{x^j y^k} - P P_{y^k} \right\} y^i.
\end{aligned}
$$

We obtain

$$R^i{}_k = \Xi \, \delta^i_k + \tau_k \, y^i, \tag{6.4}$$

where

$$\Xi := P^2 - P_{x^k} y^k, \qquad \tau_k = 3(P_{x^k} - P P_{y^k}) + \Xi_{y^k}.$$

We have

$$\tau_k y^k = -\Xi \tag{6.5}$$

and

$$R_{jk} := g_{ij} R^i{}_k = \Xi g_{jk} + \tau_k g_{ij} y^i.$$

By (2.44), $R_{jk} = R_{kj}$. This implies

$$\tau_k g_{ij} y^i = \tau_j g_{ik} y^i. \tag{6.6}$$

Contracting (6.6) with y^j and using (6.5) yields

$$\tau_k F^2 = -\Xi g_{ik} y^i.$$

Namely, $\tau_k = -\Xi F^{-1} F_{y^k}$. Then we obtain

$$R^i_{\ k} = \Xi \delta^i_k + \tau_k y^i = \Xi \left\{ \delta^i_k - F^{-1} F_{y^k} y^i \right\}. \tag{6.7}$$

Thus F is of scalar flag curvature and its flag curvature is given by

$$\mathbf{K} = \frac{\Xi}{F^2} = \frac{P^2 - P_{x^k} y^k}{F^2}. \tag{6.8}$$

Q.E.D.

Example 6.1.4 Consider the following family of Riemannian metrics:

$$\alpha_\mu := \frac{\sqrt{|y|^2 + \mu(|x|^2 |y|^2 - \langle x, y \rangle^2)}}{1 + \mu |x|^2}, \qquad y \in T_x \mathrm{B}^n(r_\mu) \cong \mathrm{R}^n.$$

According to Example 3.4.2, the spray coefficients $G^i = P y^i$, where

$$P = -\frac{\mu \langle x, y \rangle}{1 + \mu |x|^2}.$$

By a direct computation, we get

$$P_{x^k} y^k = -\mu \frac{|y|^2(1 + \mu |x|^2) - 2\mu \langle x, y \rangle^2}{(1 + \mu |x|^2)^2}.$$

By (6.8), we obtain that $\mathbf{K} = \mu$. Thus α_μ has constant sectional curvature.

In Riemannian geometry, E. Cartan's local classification theorem asserts that any Riemannian metric of constant sectional curvature μ is locally isometric to α_μ.

Example 6.1.5 Consider the following Randers metric on the unit ball $\mathrm{B}^n(1) \subset \mathrm{R}^n$,

$$F = \frac{\sqrt{|y|^2 - (|x|^2 |y|^2 - \langle x, y \rangle^2)} + \langle x, y \rangle}{1 - |x|^2} + \frac{\langle a, y \rangle}{1 + \langle a, x \rangle}, \tag{6.9}$$

where $a \in \mathbf{R}^n$ with $|a| < 1$. By Example 5.2.1, the spray coefficients of F are in the form $G^i = Py^i$, where

$$P = \frac{1}{2} \left\{ \frac{\sqrt{|y|^2 - (|x|^2|y|^2 - \langle x, y \rangle^2)} + \langle x, y \rangle}{1 - |x|^2} - \frac{\langle a, y \rangle}{1 + \langle a, x \rangle} \right\}.$$

Thus, F is projectively flat. By a direct computation,

$$P_{x^k} y^k = \frac{1}{2} \left\{ \frac{(\sqrt{|y|^2 - (|x|^2|y|^2 - \langle x, y \rangle^2)} + \langle x, y \rangle)^2}{(1 - |x|^2)^2} + \frac{\langle a, y \rangle^2}{(1 + \langle a, x \rangle)^2} \right\}.$$

By (6.8) we obtain that $\mathbf{K} = -1/4$. See [88] for more details.

The Finsler metric in (6.9) is a special Randers metric. Let us take a look at more general Randers metrics. Let $F = \alpha + \beta$ be an n-dimensional Randers metric on a manifold M, where $\alpha = \sqrt{a_{ij}(x) y^i y^j}$ is a Riemannian metric and $\beta = b_i(x) y^i$ is a closed 1-form. We shall continue to use the same abbreviations as in Lemma 3.1.2. The spray coefficients G^i of F and the spray coefficients G^i_α of α are related by $G^i = G^i_\alpha + Py^i + Q^i$, where

$$P := \frac{e_{00}}{2F} - s_0, \qquad Q^i = \alpha s^i{}_0.$$

In general, the expression for the Riemann curvature is very complicated. Thus we assume that the 1-form β is closed. Then $s_{ij} = 0$ and $s_i = 0$. In this case, $G^i = G^i_\alpha + Py^i$, where

$$P = \frac{e_{00}}{2F} - s_0 = \frac{r_{00}}{2F}.$$

Substituting $G^i = G^i_\alpha + Py^i$ into (2.49) yields

$$R^i{}_k = \bar{R}^i{}_k + \left(3 \left(\frac{\Phi}{2F} \right)^2 - \frac{\Psi}{2F} \right) \left\{ \delta^i_k - \frac{F_{y^k}}{F} y^i \right\} + \tau_k \, y^i, \qquad (6.10)$$

where

$$\Phi := b_{i;j} y^i y^j, \qquad \Psi := b_{i;j;k} y^i y^j y^k, \qquad \tau_k = \frac{1}{F} \left(b_{i;j;k} - b_{i;k;j} \right) y^i y^j. \quad (6.11)$$

We can use the above formula to calculate the Riemann curvature when β is closed.

Example 6.1.6   ([24]) Let (S^n, α) be the standard unit sphere in $\mathbf{R}^{n+1}$ and $f(x) := \varepsilon x^i$, where ε is an arbitrary nonnegative number and x^i is one of the position functions $\varphi = (x^1, \cdots, x^{n+1}) : S^n \to R^{n+1}$. $f(x)$ is an eigenfunction corresponding to the first eigenvalue $\lambda_1 = n$. In fact, it satisfies a stronger PDE:

$$f_{;i;j} = -\delta_{ij} f, \tag{6.12}$$

where the covariant derivatives of f are taken with respect to an orthonormal frame of α. It follows from (6.12) that $\delta := f(x)^2 + |df_x|^2$ is a constant. We choose a small ε such that $\delta < 1$. Define

$$F := \alpha(x, y) + \beta(x, y),$$

where $\beta = b_i(x) y^i$ is given by

$$b_i = -\frac{f_{;i}(x)}{\sqrt{1 - f(x)^2}}.$$

Note that β is closed. According to Example 3.3.2, F is projectively equivalent to α. Since α is locally projectively flat, F must be locally projectively flat. Moreover, the geodesics of F are great circles. Since $\beta = d\varphi$ is the differential of the scalar function, $\varphi = -\arcsin(f(x))$, the great circles have F-length of 2π. To see this, take an arbitrary great circle C parametrized by $c = c(t)$, $0 \leq t \leq 2\pi$, where t is the arclength parameter with respect to α. Then

$$F(c(t), \dot{c}(t)) = 1 + \varphi'(t),$$

where $\varphi(t) := \varphi(c(t))$. Then

$$\mathcal{L}_F(C) = \int_0^{2\pi} F(c(t), \dot{c}(t)) dt = \int_0^{2\pi} \left(1 + \varphi'(t)\right) dt = 2\pi.$$

This proves our claim.

Now let us compute the S-curvature. Let

$$\rho := \ln \sqrt{1 - \|\beta_x\|_\alpha^2} = \ln \sqrt{\frac{1 - f(x)^2 - |df_x|^2}{1 - f(x)^2}}.$$

By (6.12), we obtain

$$\rho_0 := \rho_{x^i} y^i = -\frac{f}{\sqrt{1 - f^2}} \beta.$$

On the other hand,

$$b_{i;j} = \frac{f}{\sqrt{1-f^2}}\Big\{\delta_{ij} - \frac{f_{;i}f_{;j}}{1-f^2}\Big\},$$

$$b_{i;j;k} = \frac{f_{;k}\delta_{ij} + f^2(f_{;j}\delta_{ik} + f_{;i}\delta_{jk})}{(1-f^2)^{3/2}} - \frac{(1+2f^2)f_{;i}f_{;j}f_{;k}}{(1-f^2)^{5/2}}. \tag{6.13}$$

This gives

$$e_{00} = \frac{f}{\sqrt{1-f^2}}\Big(\alpha^2 - \beta^2\Big).$$

By (5.10), we obtain

$$\mathbf{S} = (n+1)\Big\{\frac{e_{00}}{2F} - \rho_0\Big\} = (n+1)\frac{f}{2\sqrt{1-f^2}}\,F.$$

Thus the S-curvature is isotropic. Since F is projectively flat, by Proposition 6.1.3, it is of scalar flag curvature. By (6.10), the flag curvature is given by

$$\mathbf{K} = \frac{1}{F^2}\Big\{\alpha^2 + 3\Big(\frac{\Phi}{2F}\Big)^2 - \frac{\Psi}{2F}\Big\}. \tag{6.14}$$

Since β is closed, by (6.13), the following holds

$$\Phi = e_{00} = \frac{f}{\sqrt{1-f^2}}\Big(\alpha^2 - \beta^2\Big),$$

and

$$\Psi = -\frac{1+2f^2}{1-f^2}\Big(\alpha^2 - \beta^2\Big)\beta.$$

Substituting them into (6.14) yields

$$\mathbf{K} = \Big\{\alpha^2 + \frac{3f^2}{4(1-f^2)}(\alpha - \beta)^2 + \frac{1+2f^2}{2(1-f^2)}(\alpha - \beta)\beta\Big\}\Big/F^2$$

$$= \frac{3}{4(1-f(x)^2)}\cdot\frac{F(x,-y)}{F(x,y)} + \frac{1}{4}.$$

We see that the flag curvature always satisfies the lower bound $\mathbf{K} > \frac{1}{4}$.

Finally, let us take a look at the Hilbert metric.

Example 6.1.7 Let $H = H(x, y)$ be the Hilbert metric on a strongly convex domain $\mathcal{U} \subset \mathbb{R}^n$,

$$H := \frac{1}{2}\left\{\Theta + \bar{\Theta}\right\},$$

where $\Theta = \Theta(x, y)$ is the Funk metric on $\mathcal{U}$ and $\bar{\Theta} := \Theta(x, -y)$. Θ satisfies (1.38) and $\bar{\Theta}$ satisfies (3.21). The Hilbert metric is a reversible Finsler metric. Usually H can't be expressed in terms of elementary functions unless $\mathcal{U}$ is defined by a Randers norm. According to Example 3.4.5, H is projectively flat and its projective factor P is given by

$$P = \frac{1}{2}\left\{\Theta - \bar{\Theta}\right\}.$$

Substituting the above expression for P into (6.8) and using (1.38), one obtains that

$$\begin{aligned}
\mathbf{K} &= \frac{(\Theta - \bar{\Theta})^2 - 2(\Theta_{x^k} - \bar{\Theta}_{x^k})y^k}{(\Theta + \bar{\Theta})^2} \\
&= \frac{(\Theta - \bar{\Theta})^2 - ([\Theta^2]_{y^k} + [\bar{\Theta}^2]_{y^k})y^k}{(\Theta + \bar{\Theta})^2} \\
&= \frac{(\Theta - \bar{\Theta})^2 - 2(\Theta^2 + \bar{\Theta}^2)}{(\Theta + \bar{\Theta})^2} = -1.
\end{aligned}$$

Thus H has constant flag curvature. The Hilbert metric on the unit ball $B^n(1) \subset \mathbb{R}^n$ is just the Klein metric discussed above.

It is also an interesting problem to study Finsler metrics with $\mathbf{K}(P, y) = \mathbf{K}(P)$ independent of $y \in P$ for every tangent plane $P \subset T_x M$. Obviously, Finsler metrics with constant flag curvature have this curvature property. There are other Finsler metrics having this curvature property. Further investigation is needed.

6.2 Second Variation of a Geodesic

The geometric meaning of the Riemann curvature lies in the second variations of a geodesic. There are two types of variations for a geodesic σ: (i) the variation of the length function of curves in a neighborhood of σ (fixing the endpoints) and (ii) the variation of geodesics in a neighborhood of σ.

First we consider the variation of the length function. Let (M, F) be a Finsler manifold and let $\sigma = \sigma(t)$, $a \le t \le b$, be a geodesic in M. Let $H : [a, b] \times (-\varepsilon, \varepsilon) \to M$ be a C^∞ variation of σ,

$$H(t, 0) = \sigma(t), \qquad a \le t \le b.$$

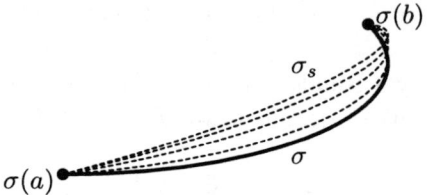

Figure 6.2

Let $\mathcal{L}(s)$ denote the length function of the curves $\sigma_s(t) := H(t, s)$, $a \le t \le b$,

$$\mathcal{L}(s) := \int_a^b F\Big(\sigma_s(t), \dot{\sigma}_s(t)\Big)dt.$$

Assuming that H fixes the endpoints,

$$H(a, 0) = \sigma(a), \qquad H(b, s) = \sigma(b), \qquad |s| < \varepsilon.$$

Let

$$V(t) := \frac{\partial H}{\partial s}(t, 0)$$

and $V^\perp(t)$ denote the orthogonal complement of $V(t)$ with respect to $\mathbf{g}_{\dot{\sigma}(t)}$,

$$V^\perp(t) := V(t) - \mathbf{g}_{\dot{\sigma}(t)}\Big(\dot{\sigma}(t), V(t)\Big).$$

Since σ is a geodesic, $\mathcal{L}'(0) = 0$ (see Section 3.2). By a direct computation, we obtain

$$\mathcal{L}''(0) = \int_a^b \Big\{ \mathbf{g}_{\dot{\sigma}(t)}\Big(D_{\dot{\sigma}}V^\perp(t), D_{\dot{\sigma}}V^\perp(t)\Big)$$
$$-\mathbf{g}_{\dot{\sigma}(t)}\Big(\mathbf{R}_{\dot{\sigma}(t)}(V^\perp(t)), V^\perp(t)\Big)\Big\}dt, \qquad (6.15)$$

where $D_{\dot{\sigma}}V^\perp(t)$ denotes the linear covariant derivative of $V^\perp(t)$ along σ. By (6.15), if the flag curvature is negative and $V^\perp(t) \not\equiv 0$, then $\mathcal{L}''(0) > 0$ and hence any geodesic σ has minimal length among curves in its neighborhood with the same endpoints.

Now we discuss another type of variation of a geodesic: the geodesic variation. Let $\sigma = \sigma(t)$, $a \leq t \leq b$, be a geodesic in a Finsler manifold (M, F). Let $H = H(s, t)$ be a variation of σ such that each curve $\sigma_s(t) := H(t, s)$, $a \leq t \leq b$, is a geodesic. Such a variation is said to be *geodesic*. Let

$$J(t) := \frac{\partial H}{\partial s}(t, 0).$$

We are going to show that $J(t)$ satisfies

$$D_{\dot{\sigma}} D_{\dot{\sigma}} J(t) + \mathbf{R}_{\dot{\sigma}(t)}(J(t)) = 0. \tag{6.16}$$

By assumption, each σ_s is a geodesic. Thus

$$\frac{\partial^2 H^i}{\partial t^2} + 2G^i\left(H, \frac{\partial H}{\partial t}\right) = 0. \tag{6.17}$$

For simplicity, let

$$T = T^i \frac{\partial}{\partial x^i} := \frac{\partial H}{\partial t}, \qquad U = U^i \frac{\partial}{\partial x^i} := \frac{\partial H}{\partial s}.$$

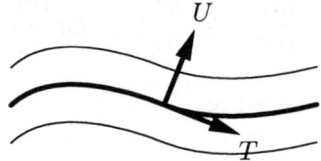

Figure 6.3

Equation (6.17) becomes

$$\frac{\partial T^i}{\partial t} + 2G^i(H, T) = 0. \tag{6.18}$$

Observe that

$$\frac{\partial T^i}{\partial s} = \frac{\partial^2 H^i}{\partial s \partial t} = \frac{\partial U^i}{\partial t}.$$

and

$$\frac{\partial}{\partial s}\Big[G^i(H,T)\Big] = U^k\frac{\partial G^i}{\partial x^k}(H,T) + \frac{\partial U^j}{\partial t}N_j^i(H,T),$$

$$\frac{\partial}{\partial t}\Big[N_j^i(H,T)\Big] = T^k\frac{\partial N_j^i}{\partial x^k}(H,T) + \frac{\partial T^k}{\partial t}\frac{\partial N_j^i}{\partial y^k}(H,T)$$

$$= T^k\frac{\partial N_j^i}{\partial x^k}(H,T) - 2G^k(H,T)\frac{\partial N_j^i}{\partial y^k}(H,T). \quad (6.19)$$

Differentiating (6.18) with respect to t yields

$$\frac{\partial^2 U^i}{\partial t^2} = -2U^k\frac{\partial G^i}{\partial x^k}(H,T) - 2\frac{\partial U^j}{\partial t}N_j^i(H,T).$$

Using the above identities, one obtains

$$D_T D_T U = D_T\Big[\Big(\frac{\partial U^i}{\partial t} + U^j N_j^i(H,T)\Big)\frac{\partial}{\partial x^i}\Big]$$

$$= -U^k\Big\{2\frac{\partial G^i}{\partial x^k} - T^j\frac{\partial N_k^i}{\partial x^j} + 2G^j\frac{\partial N_j^i}{\partial y^k} - N_j^i N_k^j\Big\}\frac{\partial}{\partial x^i}. \quad (6.20)$$

One can express the Riemann tensor $R^i{}_k$ defined in (2.49) as follows

$$R^i{}_k := 2\frac{\partial G^i}{\partial x^k} - y^j\frac{\partial N_k^i}{\partial x^j} + 2G^j\frac{\partial N_j^i}{\partial y^k} - N_j^i N_k^j.$$

Thus (6.20) can be expressed as

$$D_T D_T U + \mathbf{R}_T(U) = 0. \quad (6.21)$$

Equation (6.21) restricted to $\sigma = \sigma_0$ is just (6.16). The above proof is due to [57], [58]. Any vector field $J(t)$ satisfying (6.16) is called a *Jacobi field* along σ. Jacobi fields play an important role in Finsler geometry [87].

For a geodesic $\sigma = \sigma(t), t \geq 0$, issuing from a point $\sigma(0) = x$ in with $y = \dot{\sigma}(0)$, one can take the following geodesic variation:

$$H(t,s) := \exp_x\Big[t(y + sv)\Big], \qquad t \geq 0, \ |s| < \varepsilon.$$

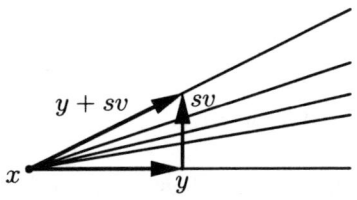

Figure 6.4

The variation field

$$J(t) := \frac{\partial H}{\partial s}(t,0) = d(\exp_x)|_{ty}(tv)$$

is a Jacobi field along σ. Under certain condition on the flag curvature or the Ricci curvature, one can estimate the zeros of $J(t)$, hence the singularity of $d\exp_x$ on T_xM.

The sphere theorem is a classical global result in Riemann-Finsler geometry. It states that a simply-connected and closed manifold with a Riemannian metric satisfying $1/4 < \mathbf{K} \le 1$ is homeomorphic to the n-sphere [54]. P. Dazord extended this theorem to *reversible* Finsler manifolds [32] [33] (see also [87]).

For a general Finsler metric F on a manifold M, let

$$\lambda := \sup_{(x,y) \in TM_o} \frac{F(x,-y)}{F(x,y)}.$$

Obviously $\lambda \ge 1$ and $\lambda = 1$ if and only if F is reversible. Thus the number λ is called the *reversibility* of F by H.B. Rademacher. Recently, he proves the following sphere theorem for (not necessary reversible) Finsler manifolds.

Theorem 6.2.1   ([78]) *Let (M,F) be a simply-connected, closed manifold of dimension $n \ge 3$ with reversibility λ. Suppose the flag curvature satisfies*

$$\left(1 - \frac{1}{1+\lambda}\right)^2 < \mathbf{K} \le 1.$$

Then M is homotopy equivalent to the n-sphere.

The sphere theorem is proved earlier for reversible Finsler manifolds in [87]. Rademacher has overcome the irreversibility obstruction. He first proves that under the condition in Theorem 6.2.1, the length of any closed geodesic is at least $\pi(1 + \frac{1}{\lambda})$. Then the theorem follows from a Rauch comparison argument and the Morse theory of the energy functional on the free loop space. We will not go into the technical part in this monograph.

6.3 Nonpositive Flag Curvature

The sign of the Riemann curvature has great implication on the geometry and topology of the manifold. In this section we are going to prove the Cartan-Hadamard theorem for nonpositively curved Finsler manifolds and a metric rigidity theorem for nonpositively curved Finsler manifolds with constant S-curvature.

Let (M, F) be a Finsler manifold. F is said to have *nonpositive flag curvature* if $\mathbf{K} \leq 0$. It is said to have *negative flag curvature* if $\mathbf{K} < 0$. The Riemann curvature is a family of self-adjoint linear maps $\mathbf{R}_y : T_x M \to T_x M$ with respect to $\mathbf{g}_y$. By (6.3), one can see that F has nonpositive flag curvature $\mathbf{K} \leq 0$ if and only if for any non-zero vectors $y, v \in T_x M$,

$$\mathbf{g}_y\Big(\mathbf{R}_y(v),\ v\Big) \leq 0.$$

Since $\mathbf{R}_y(y) = 0$ for any $y \in T_x M \setminus \{0\}$, F has negative flag curvature $\mathbf{K} < 0$ if and only if for any non-zero vectors $y, v \in T_x M$ with $\mathbf{g}_y(y, v) = 0$,

$$\mathbf{g}_y\Big(\mathbf{R}_y(v),\ v\Big) < 0.$$

Assume that (M, F) is positively complete. Then at any point $x \in M$, the exponential map $\exp_x : T_x M \to M$ is defined on the whole tangent space. For non-zero vectors $y, v \in T_x M \setminus \{0\}$, let

$$H(t, s) := \exp_x[t(y + sv)], \qquad 0 \leq t < \infty,\ |s| < \varepsilon.$$

$H = H(t, s)$ is a geodesic variation of the geodesic $\sigma(t) := \exp_x(ty)$. Then

$$J(t) := \frac{\partial H}{\partial s}(t, 0) = d(\exp_x)|_{ty}(tv), \qquad 0 \leq t < \infty, \qquad (6.22)$$

is a Jacobi field along σ, that is, it satisfies (6.16). $J = J(t)$ is C^∞ at $t = 0$ with

$$J(0) = 0, \qquad D_{\dot\sigma} J(0) = v. \qquad (6.23)$$

Conversely, for any $y, v \in T_x M \setminus \{0\}$, the Jacobi field $J = J(t)$ along the geodesic $\sigma = \exp_x(ty)$ with (6.23) is given by (6.22).

From (6.22), one can see that $\exp_x$ is singular at $ry \in T_xM$, where $r > 0$ and $y \neq 0$, if and only if there is a non-zero Jacobi field $J = J(t)$ along the geodesic $\sigma := \exp_x(ty)$, $0 \leq t \leq r$, with $J(0) = 0 = J(r)$. Thus, to prove the regularity of $\exp_x$ at ry, it suffices to prove that any Jacobi field $J(t)$, $t \geq 0$, with $J(0) = 0$ and $D_{\dot{\sigma}}J(0) \neq 0$, does not vanish at $t = r$.

Theorem 6.3.1 (Cartan-Hadamard [3]) *Let (M, F) be a positively complete Finsler manifold. Suppose that the flag curvature $\mathbf{K} \leq 0$. Then for any point $x \in M$, the exponential map $\exp_x : T_xM \to M$ is non-singular.*

Proof: Let $y, v \in T_xM \setminus \{0\}$ and let $J(t)$ be the Jacobi field along the geodesic $\sigma = \exp_x(ty)$ satisfying (6.23). Let

$$f(t) := \mathbf{g}_{\dot{\sigma}(t)}\Big(J(t),\ J(t)\Big), \qquad t \geq 0.$$

By (6.16), one obtains

$$\frac{1}{2}f'(t) = \mathbf{g}_{\dot{\sigma}}\Big(D_{\dot{\sigma}}J(t), J(t)\Big)$$

$$= \int_0^t \frac{d}{dt}\Big[\mathbf{g}_{\dot{\sigma}}\Big(D_{\dot{\sigma}}J, J\Big)\Big]d\tau$$

$$= \int_0^t \Big\{\mathbf{g}_{\dot{\sigma}}\Big(D_{\dot{\sigma}}J, D_{\dot{\sigma}}J\Big) + \mathbf{g}_{\dot{\sigma}}\Big(D_{\dot{\sigma}}D_{\dot{\sigma}}J, J\Big)\Big\}d\tau$$

$$= \int_0^t \Big\{\mathbf{g}_{\dot{\sigma}}\Big(D_{\dot{\sigma}}J, D_{\dot{\sigma}}J\Big) - \mathbf{g}_{\dot{\sigma}}\Big(\mathbf{R}_{\dot{\sigma}}(J), J\Big)\Big\}d\tau \geq 0.$$

This implies that $f(t)$ is non-decreasing. Note that

$$\frac{1}{2}f''(0) = \mathbf{g}_y(v, v) > 0.$$

One concludes that $f(t) > 0$ for all $t > 0$. Thus $\exp_x$ is non-singular at $ry \in T_xM$ for any $r > 0$.

Q.E.D.

By definition, an n-dimensional Finsler metric F has isotropic S-curvature if $\mathbf{S} = (n + 1)cF$ for some scalar function $c = c(x)$ on M, and it has constant S-curvature if $c(x) = constant$. Note that for reversible Finsler metrics, $F(x, -y) = F(x, y)$, the S-curvature $\mathbf{S}$ is isotropic if and only if $\mathbf{S} = 0$. However, there are lots of irreversible Finsler metrics with non-zero isotropic S-curvature. See Examples 5.1.3 and 6.1.6 above. Here we are going to prove a rigidity theorem for complete Finsler manifolds with nonpositive flag curvature and constant S-curvature.

Theorem 6.3.2 ([92]) *Let (M, F) be a complete Finsler manifold with nonpositive flag curvature. Suppose that F has constant S-curvature and bounded mean Cartan torsion. Then F is weakly Landsbergian ($\mathbf{J} = 0$) with $\mathbf{R}_y(\mathbf{I}_y) = 0$. Moreover, F is Riemannian at points where the flag curvature is negative.*

Proof. Let $\{\mathbf{b}_i\}$ be a local frame for TM and $\{\mathbf{e}_i := (x, y, \mathbf{b}_i)\}$ be the corresponding local frame for π^*TM. Let $\{\theta^i\}$ be dual to $\{\mathbf{b}_i\}$ and $\{\omega^i := \pi^*\theta^i\}$ be dual to $\{\mathbf{e}_i\}$. The mean Cartan tensor $\mathcal{I} = I_i\omega^i$ can be expressed as a family of vectors, $\mathbf{I}_y = I^i(x, y)\mathbf{b}_i$, where $I^i := g^{ij}I_j$, and the Landsberg tensor $\mathcal{J} = J_i\omega^i$ can also be expressed as a family of vectors, $\mathbf{J}_y = J^i(x, y)\mathbf{b}_i$, where $J^i := g^{ij}J_j$. Thus

$$\mathbf{R}_y(\mathbf{I}_y) = R^i{}_m I^m \mathbf{b}_i.$$

By assumption, $\mathbf{S} = (n + 1)cF$ for some constant c. Thus

$$\mathbf{S}_{\cdot k|m}y^m - \mathbf{S}_{|k} = (n + 1)c\left\{F_{\cdot k|m}y^m - F_{|k}\right\} = 0.$$

It follows from (5.41) that

$$J^i{}_{|m}y^m + R^i{}_m I^m = 0. \tag{6.24}$$

Let $\sigma = \sigma(t)$ be an arbitrary geodesic. Since F is complete, one may assume that σ is defined on $(-\infty, \infty)$. Let

$$\mathbf{I}(t) := I^k(\sigma(t), \dot\sigma(t))\mathbf{b}_k|_{\sigma(t)}, \qquad \mathbf{J}(t) := J^k(\sigma(t), \dot\sigma(t))\mathbf{b}_k|_{\sigma(t)}.$$

By (2.68), $J^k = I^k{}_{|m}y^m$. Thus

$$\mathbf{D}_{\dot\sigma}\mathbf{I}(t) = \dot\sigma^p(t)I^k{}_{|p}(\sigma(t), \dot\sigma(t))\mathbf{b}_k|_{\sigma(t)} = \mathbf{J}(t) \tag{6.25}$$

and

$$\mathbf{D}_{\dot\sigma}\mathbf{J}(t) = \dot\sigma^p(t)J^k{}_{|p}(\sigma(t), \dot\sigma(t))\mathbf{b}_k|_{\sigma(t)} = \mathbf{D}_{\dot\sigma}\mathbf{D}_{\dot\sigma}\mathbf{I}(t). \tag{6.26}$$

Equation (6.24) restricted to $\sigma(t)$ becomes

$$\mathbf{D}_{\dot\sigma}\mathbf{D}_{\dot\sigma}\mathbf{I}(t) + \mathbf{R}_{\dot\sigma(t)}(\mathbf{I}(t)) = 0. \tag{6.27}$$

Thus, the mean Cartan torsion is a Jacobi field along any geodesic.

Let

$$\varphi(t) := \mathbf{g}_{\dot{\sigma}(t)}\Big(\mathbf{I}(t), \mathbf{I}(t)\Big).$$

It follows from (6.25), (6.26) and (6.27) that

$$\varphi''(t) = 2\mathbf{g}_{\dot{\sigma}(t)}\Big(D_{\dot{\sigma}}D_{\dot{\sigma}}\mathbf{I}(t), \mathbf{I}(t)\Big) + 2\mathbf{g}_{\dot{\sigma}(t)}\Big(D_{\dot{\sigma}}\mathbf{I}(t), D_{\dot{\sigma}}\mathbf{I}(t)\Big)$$

$$= -2\mathbf{g}_{\dot{\sigma}(t)}\Big(\mathbf{R}_{\dot{\sigma}(t)}(\mathbf{I}(t)), \mathbf{I}(t)\Big) + 2\mathbf{g}_{\dot{\sigma}(t)}\Big(\mathbf{J}(t), \mathbf{J}(t)\Big). \qquad (6.28)$$

By assumption, $\mathbf{K} \leq 0$. It follows from (6.28) that

$$\varphi''(t) \geq 0.$$

Thus $\varphi(t)$ is convex and nonpositive. Suppose that $\varphi'(t_o) \neq 0$ for some t_o. If $\varphi'(t_o) < 0$, then

$$\varphi(t) \geq \varphi(t_o) - \varphi'(t_o)(t_o - t), \qquad t < t_o.$$

If $\varphi'(t_o) > 0$, then

$$\varphi(t) \geq \varphi(t_o) + \varphi'(t_o)(t - t_o), \qquad t > t_o.$$

One can see that $\lim_{t \to +\infty} \varphi(t) = \infty$ or $\lim_{t \to -\infty} \varphi(t) = \infty$. This implies that the mean Cartan torsion is unbounded, that contradicts the assumption. Therefore, $\varphi'(t) = 0$ and hence $\varphi''(t) = 0$. It follows from (6.28) that

$$\mathbf{R}_{\dot{\sigma}(t)}(\mathbf{I}(t)) = 0, \qquad \mathbf{J}(t) = 0.$$

Since σ is arbitrary, one can conclude that

$$\mathbf{R}_y(\mathbf{I}_y) = 0, \qquad \mathbf{J}_y = 0.$$

Assume that F has negative flag curvature at a point $x \in M$. Since the vector $\mathbf{I}_y$ is orthogonal to y with respect to $\mathbf{g}_y$, and $\mathbf{R}_y(\mathbf{I}_y) = 0$, one concludes that $\mathbf{I}_y = 0$ for all $y \in T_x M \setminus \{0\}$. By Deicke's theorem (Theorem 1.5.1), F is Riemannian.   Q.E.D.

Any Finsler metric on a closed manifold is complete with bounded Cartan torsion. One immediately obtains the following

Corollary 6.3.3 *Let (M, F) be a closed Finsler manifold of negative flag curvature. If F has constant S-curvature, then it must be Riemannian.*

It is not difficult to show that every complete Finsler surface with non-positive Gauss curvature, constant S-curvature and bounded mean Cartan torsion is either Riemannian or locally Minkowskian. First one sees that such a Finsler surface must be weakly Landsbergian ($\mathbf{J} = 0$). Then the conclusion follows from a global rigidity theorem in [6]. The details are left to the reader for an exercise.

Example 6.3.4 Let $f : [0, \infty) \times [0, \infty) \to [0, \infty)$ be an arbitrary C^∞ function satisfying (1.18), (1.20) and (1.21). Let (M_i, α_i), $i = 1, 2$, be arbitrary Riemannian manifolds and $M = M_1 \times M_2$. Let

$$F := \sqrt{f\Big([\alpha_1(x_1, y_1)]^2, \ [\alpha_2(x_2, y_2)]^2\Big)},$$

where $x = (x_1, x_2) \in M$ and $y = y_1 \oplus y_2 \in T_{(x_1, x_2)}(M_1 \times M_2) \cong T_{x_1}M_1 \oplus T_{x_2}M_2$. By Example 1.2.5, F is a Finsler metric. By Example 4.3.1, the spray coefficients of F are given by

$$G^a(x, y) = \bar{G}^a(x_1, y_1), \qquad G^\alpha(x, y) = \bar{G}^\alpha(x_1, y_1), \qquad (6.29)$$

where $\bar{G}^a$ and $\bar{G}^\alpha$ are the spray coefficients of α_1 and α_2 respectively. From (6.29), we can see that F is a Berwald metric. Thus by Proposition 5.1.2, $\mathbf{S} = 0$.

By a direct computation, one obtains the following formula for the Riemann tensor of F:

$$\left(R^i{}_j \right) = \left(\bar{R}^i{}_j \right) = \begin{pmatrix} \bar{R}^a{}_b & 0 \\ 0 & \bar{R}^\alpha{}_\beta \end{pmatrix},$$

where $\bar{R}^a{}_b$ and $\bar{R}^\alpha{}_\beta$ are the coefficients of the Riemann tensor of α_1 and α_2 respectively. Let $R_{ij} := g_{ik}R^k{}_j$, $\bar{R}_{ab} := \bar{g}_{ac}\bar{R}^c{}_b$ and $\bar{R}_{\alpha\beta} := \bar{g}_{\alpha\gamma}\bar{R}^\gamma{}_\beta$. Using (1.19), one obtains

$$\left(R_{ij} \right) = \begin{pmatrix} f_s\bar{R}_{ab} & 0 \\ 0 & f_t\bar{R}_{\alpha\beta} \end{pmatrix}.$$

For any vector $v = v^i \frac{\partial}{\partial x^i}\big|_x \in T_xM$,

$$\mathbf{g}_y\Big(\mathbf{R}_y(v), v\Big) = f_s\bar{R}_{ab}v^a v^b + f_t\bar{R}_{\alpha\beta}v^\alpha v^\beta. \qquad (6.30)$$

Assume that α_1 and α_2 both have nonpositive sectional curvature. Then it follows from (6.30) that F has nonpositive flag curvature.

Using (1.22), we can compute the mean Cartan torsion. First, observe that

$$I_i = \frac{\partial}{\partial y^i}\left[\ln\sqrt{\det(g_{jk})}\right] = \frac{\partial}{\partial y^i}\left[\ln\sqrt{h\left([\alpha_1]^2, [\alpha_2]^2\right)}\right].$$

We obtain

$$I_a = \frac{h_s}{h}\bar{y}_a \qquad I_\alpha = \frac{h_t}{h}\bar{y}_\alpha,$$

where $\bar{y}_a := \bar{g}_{ab}y^b$ and $\bar{y}_\alpha := \bar{g}_{\alpha\beta}y^\beta$. Clearly, the mean Cartan torsion is bounded.

Since $\bar{y}_a\bar{R}^a{}_b = 0$ and $\bar{y}_\alpha\bar{R}^\alpha{}_\beta = 0$, we have

$$\mathbf{g}_y\left(\mathbf{R}_y(\mathbf{I}_y), \mathbf{I}_y\right) = I_i R^i{}_j I^j = \frac{h_s}{h}\bar{y}_a\bar{R}^a{}_b I^b + \frac{h_t}{h}\bar{y}_\alpha\bar{R}^\alpha{}_\beta I^\beta = 0.$$

Thus the Riemann curvature vanishes on the mean Cartan torsion.

Since F is a Berwald metric, by Proposition 2.1.3, it is a Landsberg metric, i.e., $\mathbf{J} = 0$. Thus the conclusion in Theorem 6.3.2 holds.

The completeness in Theorem 6.3.2 can't be dropped. See Example 6.1.5 and the following *Fish tank* metric.

Example 6.3.5 ([90]) Let $n \geq 2$ and

$$\mathcal{U} := \left\{p = (s, t, \bar{p}) \in \mathrm{R}^2 \times \mathrm{R}^{n-2} \ \middle|\ s^2 + t^2 < 1\right\}.$$

Define $F = F(x, y)$

$$F(x, y) := \frac{\sqrt{\left(-tu + sv\right)^2 + |y|^2\left(1 - s^2 - t^2\right)} - \left(-tu + sv\right)}{1 - s^2 - t^2},$$

where $y = (u, v, \bar{y}) \in T_x\mathcal{U} \cong \mathrm{R}^n$ and $x = (s, t, \bar{p}) \in \Omega$. One can verify that F is a Finsler metric on Ω with vanishing flag curvature $\mathbf{K} = 0$ and vanishing S-curvature $\mathbf{S} = 0$. One can verify that $\mathbf{J} \neq 0$. Thus Theorem 6.3.2 does not hold if the assumption on completeness is dropped.

From the above discussion, we may ask the following question: is there any non-Berwaldian Finsler metric of dimension $n \geq 3$ satisfying the following conditions:

$$\mathbf{K} = 0, \quad \mathbf{S} = 0, \quad \mathbf{J} = 0?$$

This problem remains open. Note that in dimension two, the conditions, $\mathbf{S} = 0$ and $\mathbf{J} = 0$, imply that the metric is a Berwald metric. Berwald metrics with $\mathbf{K} = 0$ are locally Minkowskian.

Chapter 7

Finsler Metrics of Scalar Flag Curvature

In this chapter, we are going to discuss Finsler metrics of scalar (flag) curvature. We have seen that every locally projectively flat Finsler metrics are of scalar flag curvature (Proposition 6.1.3). There are Finsler metrics of scalar flag curvature which are not locally projectively flat. This shows the complexity and richness of general Finsler metrics. It is our goal to reveal the relationship between the flag curvature and other non-Riemannian quantities for Finsler metrics of scalar flag curvature.

7.1 Some Basic Properties

Let (M, F) be a Finsler manifold. Assume that F is of scalar flag curvature, that is, the flag curvature $\mathbf{K} = \mathbf{K}(x, y)$ is a scalar function on $TM \setminus \{0\}$. This curvature condition is equivalent to the following identity in any standard local coordinate system,

$$R^i{}_k = \mathbf{K}F^2 \, h^i{}_k = \mathbf{K}\Big\{F^2\delta^i_k - g_{kq}y^q y^i\Big\}, \qquad (7.1)$$

where $h^i{}_k := \delta^i_k - F^{-2}g_{kq}y^q y^i$. It follows from (2.73) and (7.1) that

$$R^i{}_{kl} = \frac{1}{3}\mathbf{K}_{\cdot l}F^2 \, h^i{}_k - \frac{1}{3}\mathbf{K}_{\cdot k}F^2 \, h^i{}_l + \mathbf{K}\Big\{g_{lp}\delta^i_k - g_{kp}\delta^i_l\Big\}y^p. \qquad (7.2)$$

To study the relationship between the flag curvature and other non-Riemannian quantities, we will need some identities. Differentiating (7.1) yields

$$R^i{}_{k \cdot l} = \mathbf{K}_{\cdot l}F^2 \, h^i{}_k + \mathbf{K}\Big\{2g_{lp}y^p\delta^m_k - g_{kp}y^p\delta^i_l - g_{kl}y^i\Big\}. \qquad (7.3)$$

127

Let $h_{ij} := g_{ik}h^k{}_j = FF_{y^i y^j}$ By (2.69), (2.70) and (7.3), one obtains

$$C_{ijk|p|q}y^p y^q = -\frac{1}{3}F^2\left\{\mathbf{K}_{\cdot i}h_{jk} + \mathbf{K}_{\cdot j}h_{ik} + \mathbf{K}_{\cdot k}h_{ij} + 3\mathbf{K}C_{ijk}\right\} \qquad (7.4)$$

and

$$I_{k|p|q}y^p y^q = -\frac{1}{3}F^2\left\{(n+1)\mathbf{K}_{\cdot k} + 3\mathbf{K}I_k\right\}. \qquad (7.5)$$

Recall the Matsumoto torsion defined by

$$M_{ijk} := C_{ijk} - \frac{1}{n+1}\left\{I_i h_{jk} + I_j h_{ik} + I_k h_{ij}\right\}.$$

It follows from (7.4) and (7.5) that

$$M_{ijk|p|q}y^p y^q + \mathbf{K}F^2 M_{ijk} = 0. \qquad (7.6)$$

This is an important equation for Finsler metrics of scalar flag curvature. By (7.6), one can show that for a Landsberg metric of scalar flag curvature on a manifold of dimension ≥ 3, if the flag curvature $\mathbf{K} \neq 0$, then it is Riemannnian ([76]).

Since F is of scalar flag curvature, (2.63) can be simplified to

$$R^i{}_{k|j} - R^i{}_{j|k} + R^i{}_{jk|m}y^m = 0. \qquad (7.7)$$

Substituting (7.1) and (7.2) into (7.7) yields

$$\mathbf{K}_{|l}h^i{}_k - \mathbf{K}_{|k}h^i{}_l - \frac{1}{3}\left\{\mathbf{K}_{\cdot l|m}h^i{}_k - \mathbf{K}_{\cdot k|m}h^i{}_l\right\}y^m$$
$$-F^{-2}\mathbf{K}_{|m}y^m\left\{g_{lp}\delta^i_k - g_{kp}\delta^i_l\right\}y^p = 0.$$

Taking the summation over $i = k$ in the above identity yields

$$(n-2)\left\{\mathbf{K}_{|l} - F^{-2}\mathbf{K}_{|m}y^m g_{lp}y^p - \frac{1}{3}\mathbf{K}_{\cdot l|m}y^m\right\} = 0. \qquad (7.8)$$

On the other hand, the equation $d^2\mathbf{K} = 0$ yields the following Ricci identity:

$$\mathbf{K}_{\cdot l|m} = \mathbf{K}_{|m\cdot l} + \mathbf{K}_{\cdot k}L^k{}_{ml}.$$

Contracting the above identity with y^m yields

$$\mathbf{K}_{\cdot l|m}y^m = \mathbf{K}_{|m\cdot l}y^m = (\mathbf{K}_{|m}y^m)_{\cdot l} - \mathbf{K}_{|l}.$$

Assume that $n \geq 3$. Substituting the above equation into (7.8) yields

$$4F^3 \mathbf{K}_{|l} - (F^3 \mathbf{K}_{|m} y^m)_{\cdot l} = 0. \tag{7.9}$$

Then we obtain the following

Lemma 7.1.1 (Schur Lemma [18]) *Let (M, F) be a Finsler manifold of dimension $n \geq 3$. Suppose that the flag curvature is isotropic, i.e., the flag curvature $\mathbf{K} = \mathbf{K}(x)$ is a function of $x \in M$ only. Then $\mathbf{K} = constant$.*

Proof: Since $\mathbf{K}_{|k} = \mathbf{K}_{x^k}(x)$, it suffices to prove that $\mathbf{K}_{|k} = 0$. Since $\mathbf{K}$ is a scalar function on M, $\mathbf{K}_{|m} = \mathbf{K}_{x^m}$ is a function of $x \in M$ only. Hence $\mathbf{K}_{|m \cdot l} = 0$. It follows from (7.9) that

$$F^2 \mathbf{K}_{|j} = (\mathbf{K}_{|m} y^m) \, g_{jp} y^p. \tag{7.10}$$

Differentiating (7.10) with respect to y^k yields

$$\mathbf{K}_{|m} y^m \, g_{jk} = 2\mathbf{K}_{|j} \, g_{kp} y^p - \mathbf{K}_{|k} \, g_{jp} y^p. \tag{7.11}$$

Let $u = u^i \frac{\partial}{\partial x^i}|_x \in T_x M$ be $\mathbf{g}_y$-orthogonal to y, namely, $\mathbf{g}_y(y, u) = 0$. Contracting (7.11) with u^j and u^k yields

$$\mathbf{K}_{|m} y^m \, \mathbf{g}_y(u, u) = 2\mathbf{K}_{|j} u^j \, \mathbf{g}_y(y, u) - \mathbf{K}_{|k} u^k \, \mathbf{g}_y(y, u) = 0.$$

Thus $\mathbf{K}_{|m} y^m = 0$. By (7.10), one concludes that $\mathbf{K}_{|j} = 0$. Q.E.D.

Many known Finsler metrics of scalar flag curvature have almost isotropic S-curvature. Thus it is natural to study Finsler metrics of scalar flag curvature with isotropic S-curvature.

Proposition 7.1.2 ([24]) *Let (M, F) be an n-dimensional Finsler manifold of scalar flag curvature with flag curvature $\mathbf{K} = \mathbf{K}(x, y)$. Suppose that the S-curvature is almost isotropic, i.e.,*

$$\mathbf{S} = (n + 1)\Big\{cF + \eta\Big\}, \tag{7.12}$$

where $c = c(x)$ is a scalar function and $\eta = \eta_i(x) y^i$ is a closed 1-form on M. Then there is a scalar function $\sigma = \sigma(x)$ on M such that the flag curvature is in the following form,

$$\mathbf{K} = 3\frac{c_{x^m} y^m}{F} + \sigma.$$

Proof. Substituting (7.3) into (5.42), one obtains

$$\mathbf{S}_{\cdot k|m} y^m - \mathbf{S}_{|k} = -\frac{n+1}{3} \mathbf{K}_{\cdot k} F^2.$$

By (7.12), one obtains

$$\mathbf{S}_{\cdot k|m} y^m - \mathbf{S}_{|k} = (n+1)\left\{ c_{|m} y^m F_{\cdot k} - c_{|k} F + \left[\eta_{k|m} - \eta_{m|k} \right] y^m \right\}$$
$$= (n+1)\left\{ c_{|m} y^m F_{\cdot k} - c_{|k} F \right\}.$$

Thus

$$c_{|m} y^m F_{\cdot k} - c_{|k} F = -\frac{1}{3} \mathbf{K}_{\cdot k} F^2. \tag{7.13}$$

Here $c = c(x)$ is viewed as a scalar function on TM_o and its covariant derivatives are defined in the usual way, i.e., $c_{|i} = c_{x^i}$ are the usual partial derivatives with respect to x^i. Rewriting (7.13) as follows

$$\left[\frac{1}{3} \mathbf{K} - \frac{c_{x^m} y^m}{F} \right]_{y^k} = 0,$$

one concludes that the following quantity

$$\sigma := \mathbf{K} - \frac{3 c_{x^m} y^m}{F}$$

is a scalar function on M. This proves the proposition.   Q.E.D.

Corollary 7.1.3   ([73]) *Let F be an n-dimensional Finsler metric of scalar flag curvature with flag curvature $\mathbf{K} = \mathbf{K}(x, y)$. If F has constant S-curvature, i.e., $\mathbf{S} = (n+1)cF$, where $c = $ constant, then $\mathbf{K} = \mathbf{K}(x)$ is a scalar function on M.*

According to Lemma 7.1.1, when $n = \dim M \geq 3$, $\mathbf{K} = \mathbf{K}(x)$ if and only if $\mathbf{K} = constant$.

We can use Proposition 7.1.2 to compute the flag curvature if the Finsler metric is of scalar flag curvature and isotropic S-curvature. First using the scalar function $c = c(x)$, one can determine the first part of the flag curvature, then choose a special vector y to determine the scalar function $\sigma = \sigma(x)$ since it is independent of y.

Example 7.1.4   For an arbitrary number ε with $0 < \varepsilon \leq 1$, let $F = \alpha + \beta$ be given by

$$\alpha := \frac{\sqrt{(1 - \varepsilon^2)(su + tv)^2 + \varepsilon(u^2 + v^2)(1 + \varepsilon(s^2 + t^2))}}{1 + \varepsilon(s^2 + t^2)},$$

$$\beta := \frac{\sqrt{1 - \varepsilon^2}(su + tv)}{1 + \varepsilon(s^2 + t^2)},$$

where $x = (s, t) \in \mathbb{R}^2$ and $y = (u, v) \in T_x\mathbb{R}^2 \cong \mathbb{R}^2$. F is a Randers metric on $\mathbb{R}^2$. This is just Example 5.2.3 in dimension two. Thus $\mathbf{S} = 3cF$, where

$$c = \frac{\sqrt{1 - \varepsilon^2}}{2(\varepsilon + s^2 + t^2)}.$$

Note that β is closed. Thus F is projectively equivalent to α. However, the Gauss curvature $\bar{\mathbf{K}}$ of α is not a constant,

$$\bar{\mathbf{K}} = \frac{2\varepsilon}{\varepsilon + s^2 + t^2} + \frac{1 - \varepsilon^2}{(\varepsilon + s^2 + t^2)^2}.$$

According to the Beltrami theorem in Riemannian geometry, α (hence F) is not locally projectively flat. Thus F is not projectively flat. Using a Maple program, we obtain the following expression for the Gauss curvature:

$$\mathbf{K} = -\frac{3\sqrt{1 - \varepsilon^2}(su + tv)}{(\varepsilon + s^2 + t^2)^2 F} + \frac{2\varepsilon}{\varepsilon + s^2 + t^2} + \frac{7(1 - \varepsilon^2)}{4(\varepsilon + s^2 + t^2)^2}.$$

7.2   Global Rigidity Theorems

In this section we are going to discuss some global rigidity properties of Finsler metrics of scalar flag curvature. We first find a special equation on the flag curvature and the Cartan torsion or the Matsumoto torsion, such that the restriction of it to an arbitrary geodesic is a second order ordinary differential equation. Then using basic comparison techniques, we show that under certain growth conditions on the Cartan torsion (resp. the Matsumoto torsion), the Cartan torsion (resp. the Matsumoto torsion) must vanish. These growth conditions are always satisfied if the manifold is closed.

Let (M, F) be a complete Finsler manifold. Let $\mathbf{M}$ be the Matsumoto torsion defined in (1.53). The norm of $\mathbf{M}$ at a point $x \in M$ is defined by

$$\|\mathbf{M}\|_x := \sup_{y,u,v,w \in T_x M \backslash \{0\}} \frac{F(x,y)|\mathbf{M}_y(u,v,w)|}{\sqrt{\mathbf{g}_y(u,u)\mathbf{g}_y(v,v)\mathbf{g}_y(w,w)}}. \tag{7.14}$$

The Matsumoto torsion grows *sub-exponentially* at rate of $k > 0$ if for any point $x \in M$

$$M(x,r) := \sup_{\min(d(z,x),d(x,z)) \leq r} \|\mathbf{M}\|_z = o(e^{kr}), \qquad (r \to +\infty).$$

The Matsumoto torsion grows *sub-linearly* if for any $x \in M$,

$$\lim_{r \to +\infty} r^{-1} M(x,r) = 0, \qquad (r \to +\infty).$$

Similarly, we define the norm of the (mean) Cartan torsion at a point $x \in M$ as follows,

$$\|\mathbf{I}\|_x := \sup_{y,u \in T_x M \backslash \{0\}} \frac{F(x,y)|\mathbf{I}_y(u)|}{\sqrt{\mathbf{g}_y(u,u)}},$$

$$\|\mathbf{C}\|_x := \sup_{y,u,v,w \in T_x M \backslash \{0\}} \frac{F(x,y)|\mathbf{C}_y(u,v,w)|}{\sqrt{\mathbf{g}_y(u,u)\mathbf{g}_y(v,v)\mathbf{g}_y(w,w)}}.$$

Let

$$I(x,r) := \sup_{\min(d(z,x),d(x,z)) \leq r} \|\mathbf{I}\|_x, \qquad C(x,r) := \sup_{\min(d(z,x),d(x,z)) \leq r} \|\mathbf{C}\|_x.$$

Then we can define the growth rate for $\mathbf{I}$ and $\mathbf{C}$ as above.

From the definition of the Matsumoto torsion in (1.52), we have

$$\|\mathbf{M}\|_x \leq \|\mathbf{C}\|_x + \frac{3}{n+1}\|\mathbf{I}\|_x \leq 3\|\mathbf{C}\|_x.$$

Thus if the Cartan torsion grows sub-exponentially at rate of k, then the Matsumoto torsion grows at the same rate. When M is closed, F is always complete and all mentioned geometric quantities are bounded.

Theorem 7.2.1 ([74]) *Let (M, F) be an n-dimensional complete Finsler manifold of scalar flag curvature with flag curvature $\mathbf{K} = \mathbf{K}(x,y) \leq -1$ ($n \geq 3$). Suppose that the Matsumoto torsion grows sub-exponentially at rate of $k = 1$. Then F is a Randers metric.*

Proof. According to Proposition 1.5.3, in dimension $n \geq 3$, a Finsler metric is a Randers metric if and only if the Matsumoto torsion vanishes. Thus it suffices to prove that the Matsumoto torsion vanishes under the assumption. Suppose that this is not true. Then $\mathbf{M}_y(u, u, u) \neq 0$ for some $y, u \in T_x M \setminus \{0\}$ with $F(x, y) = 1$. Let $\sigma = \sigma(t)$ be the unit speed geodesic with $\sigma(0) = x$ and $\dot{\sigma}(0) = y$, and let $U = U^i(t) \frac{\partial}{\partial x^i}|_{\sigma(t)}$ be the linearly parallel vector field along σ with $U(0) = u$, that is, $D_{\dot{\sigma}} U(t) = 0$. Let

$$\mathcal{M}(t) := \mathbf{M}_{\dot{\sigma}(t)}\Big(U(t), U(t), U(t)\Big) = M_{ijk}\Big(\sigma(t), \dot{\sigma}(t)\Big) U^i(t) U^j(t) U^k(t).$$

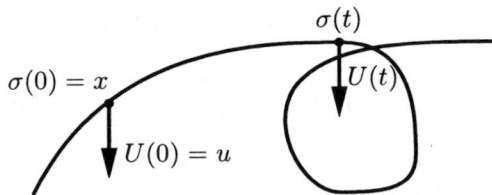

Figure 7.1

It follows from (7.6) that

$$\mathcal{M}''(t) + \mathbf{K}(t)\mathcal{M}(t) = 0, \tag{7.15}$$

where $\mathbf{K}(t) := \mathbf{K}\Big(\sigma(t), \dot{\sigma}(t)\Big) \leq -1$. We are going to estimate $\mathcal{M}(t)$ by comparing it with the following function:

$$\mathcal{M}_o(t) := \mathcal{M}(0) \cosh(t) + \mathcal{M}'(0) \sinh(t).$$

Note that $\mathcal{M}_o(t)$ satisfies

$$\mathcal{M}_o''(t) - \mathcal{M}_o(t) = 0. \tag{7.16}$$

Let (a, b) be the maximal interval on which $\mathcal{M}(t) \neq 0$. Let

$$f(t) := \frac{\mathcal{M}'(t)}{\mathcal{M}(t)}, \qquad a < t < b.$$

It follows from (7.15) that

$$f'(t) + f(t)^2 = -\mathbf{K}(t) \geq 1. \tag{7.17}$$

Let (α, β) be the maximal internal on which $\mathcal{M}_o(t) \neq 0$. Let

$$f_o(t) := \frac{\mathcal{M}_o'(t)}{\mathcal{M}_o(t)}, \qquad \alpha < t < \beta.$$

It follows from (7.16) that

$$f_o'(t) + f_o(t)^2 = 1. \tag{7.18}$$

We claim that $\varphi(t) := |\mathcal{M}(t)/\mathcal{M}_o(t)|$ attains its minimum $\varphi(0) = 1$ at $t = 0$. To show this, consider the following function,

$$h(t) := \left\{ f(t) - f_o(t) \right\} \exp \left\{ \int [f(t) + f_o(t)] dt \right\}.$$

By (7.17) and (7.18), we have

$$h'(t) = \left\{ [f'(t) + f(t)^2] - [f_o'(t) + f_o(t)^2] \right\} \exp \left\{ \int [f(t) + f_o(t)] dt \right\} \geq 0.$$

Note that $h(0) = 0$. Thus $h(t) \leq 0$ for $t < 0$ and $h(t) \geq 0$ for $t > 0$. Since $h(t)$ has the same sign as $f(t) - f_o(t)$, we conclude that

$$\frac{\varphi'(t)}{\varphi(t)} = f(t) - f_o(t) = \begin{cases} \leq 0 & \text{if } t < 0 \\ \geq 0 & \text{if } t > 0. \end{cases}$$

This implies that $\varphi'(t) \leq 0$ for $t < 0$ and $\varphi'(t) \geq 0$ for $t > 0$. Therefore, $\varphi(t)$ attains its minimum at $t = 0$. We conclude that $\varphi(t) \geq \varphi(0) = 1$ for $\max(a, \alpha) < t < \min(b, \beta)$, i.e.,

$$\left| \mathcal{M}(t) \right| \geq \left| \mathcal{M}_o(t) \right|. \tag{7.19}$$

Clearly, $(\alpha, \beta) \subset (a, b)$. Since $d(\sigma(-t), x) \leq t$ and $d(x, \sigma(t)) \leq t$ for any $t > 0$, one gets

$$M(x, r) \geq \max \left\{ \mathcal{M}(t) \mid |t| \leq r \right\}.$$

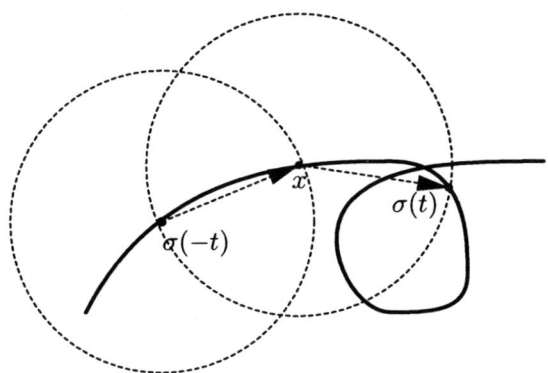

Figure 7.2

Suppose that $\mathcal{M}'(0) = 0$ or it has the same sign as $\mathcal{M}(0)$. Since $\mathcal{M}_o(t) \neq 0$ for all $t > 0$. Thus $\beta = \infty$ and

$$M(x, r) \geq \left|\mathcal{M}(r)\right| \geq \left|\mathcal{M}(0)\right| \cosh(r) + \left|\mathcal{M}'(0)\right| \sinh(r), \qquad r > 0.$$

Suppose that $\mathcal{M}'(0)$ has the opposite sign as $\mathcal{M}(0)$. Since $\mathcal{M}_o(t) \neq 0$ for all $t < 0$. Thus $\alpha = -\infty$ and

$$M(x, r) \geq \left|\mathcal{M}(-r)\right| \geq \left|\mathcal{M}(0)\right| \cosh(r) + \left|\mathcal{M}'(0)\right| \sinh(r), \qquad r > 0.$$

In either case,

$$\liminf_{r \to \infty} \frac{M(x, r)}{e^r} \geq \frac{1}{2}\left\{|\mathcal{M}(0)| + |\mathcal{M}'(0)|\right\} > 0.$$

But $M(x, r)$ grows exponentially at rate of $k = 1$. This is a contradiction. Thus the Matsumoto torsion vanishes. By Proposition 1.5.3, F must be a Randers metric. Q.E.D.

Consider a locally projectively flat complete Finsler metric F on a manifold of dimension $n \geq 3$. First, by Proposition 6.1.3, it is of scalar flag curvature. Assume that the flag curvature satisfies $\mathbf{K} \leq -1$ and the Matsumoto torsion grows sub-exponentially at rate of $k = 1$, then $F = \alpha + \beta$ is a Randers metric. Moreover, by Proposition 3.4.8, the Riemannian metric α is locally projectively flat and the 1-form β is closed. Since any Finsler metric on a *closed* manifold is complete with bounded Cartan torsion (and hence bounded Matsumoto torsion), one obtains the following

Corollary 7.2.2 *Let F be a Finsler metric on a closed manifold M of dimension $n \geq 3$. Suppose that F is of scalar flag curvature with negative flag curvature, then it is a Randers metric. In particular, if F is locally projectively flat with negative flag curvature, then it is a locally projectively flat Randers metric.*

Let α be a Riemannian metric of negative constant sectional curvature on a closed manifold M. According to Examples 3.4.2 and 6.1.4, α is locally projectively flat. Let β be an arbitrary closed 1-form on M. Then for sufficiently small ε, $F_\varepsilon := \alpha + \varepsilon\beta$ is a locally projectively flat Randers metric, since it is projectively equivalent to α by Proposition 3.4.8. Therefore F_ε is of scalar flag curvature by Proposition 6.1.3. By continuity, the flag curvature of F_ε is negative for small ε.

If we impose the reversibility condition on the Finsler mertric, we obtain the following

Corollary 7.2.3 *Let F be a reversible Finsler metric on a closed manifold M of dimension $n \geq 3$. Suppose that F is of scalar flag curvature with negative flag curvature, then it is a Riemannian metric of constant negative sectional curvature.*

The reversibility condition in Corollary 7.2.3 can be dropped if we assume that the flag curvature is isotropic, i.e., $\mathbf{K} = \mathbf{K}(x)$ is independent of directions. See Corollary 7.2.5 below. First, we prove the following

Theorem 7.2.4 *Let (M, F) be a complete Finsler manifold with isotropic flag curvature $\mathbf{K} = \mathbf{K}(x)$.*

(a) *If $\mathbf{K} \leq -1$ and $\mathbf{I}$ grows sub-exponentially at rate of $k = 1$, then F is Riemannian.*

(b) *If $\mathbf{K} \leq 0$ and $\mathbf{C}$ (resp. $\mathbf{I}$) grows sub-linearly, then F is Landsbergian (resp. weakly Landsbergian). Further F is Riemannian on any open subset where $\mathbf{K} < 0$.*

Proof: By assumption, $\mathbf{K}_{\cdot k} = 0$. It follows from (2.68) and (7.5) that

$$I_{i|p|q}y^p y^q + \mathbf{K}F^2 I_i = 0. \tag{7.20}$$

Let $y, u \in T_x M \setminus \{0\}$ be arbitrary vectors with $F(x, y) = 1$. Let $\sigma = \sigma(t)$ be the unit speed geodesic with $\sigma(0) = x$ and $\dot{\sigma}(0) = y$, and let $U = U(t)$

be the parallel vector field along σ with $U(0) = u$. Set

$$\mathcal{I}(t) := \mathbf{I}_{\dot{\sigma}(t)}\Big(U(t)\Big) = I_i\Big(\sigma(t), \dot{\sigma}(t)\Big)U^i(t).$$

It follows from (7.20) that

$$\mathcal{I}''(t) + \mathbf{K}(t)\mathcal{I}(t) = 0, \tag{7.21}$$

where $\mathbf{K}(t) := \mathbf{K}(\sigma(t))$.

Suppose that $\mathbf{K} \leq -1$ and the mean Cartan torsion $\mathbf{I}$ grows sub-exponentially at rate of $k = 1$, i.e., $I(x, r) = o(e^r)$ for any $x \in M$. By the same argument as in Theorem 7.2.1, one can show that $\mathcal{I}(0) = I_i(x, y)u^i = 0$. Since y and u are arbitrary, one concludes that the mean Cartan torsion $\mathbf{I} = 0$. Therefore F is Riemannian by Theorem 1.5.1.

Suppose that $\mathbf{K} \leq 0$ and the mean Cartan torsion grows sub-linearly. We first show that $\mathbf{J} = 0$. We prove this by contradiction. Assume that $\mathbf{J}_y(u) \neq 0$ for some $y, u \in T_x M \setminus \{0\}$. Let $\mathbf{I}(t)$ be defined as above for y, u so that $\mathbf{I}(0) = \mathbf{I}_y(u)$ and $\mathcal{I}'(0) = \mathbf{J}_y(u)$. Since $\mathcal{I}'(0) \neq 0$, $\mathcal{I}(\varepsilon) \neq 0$ and $\mathcal{I}'(\varepsilon) \neq 0$ for small $\varepsilon > 0$. Thus we may assume that $\mathcal{I}(0) \neq 0$ and $\mathcal{I}'(0) \neq 0$. We are going to show the following inequality,

$$\Big|\mathcal{I}(t)\Big| \geq \Big|\mathcal{I}(0) + \mathcal{I}'(0)t\Big|, \qquad \alpha < t < \beta, \tag{7.22}$$

where (α, β) is the maximal interval containing 0 on which the function on the right of (7.22) is not equal to zero.

Let

$$\mathcal{I}_o(t) := \mathcal{I}(0) + \mathcal{I}'(0)t.$$

Let

$$f(t) := \frac{\mathcal{I}'(t)}{\mathcal{I}(t)}, \qquad f_o(t) := \frac{\mathcal{I}'_o(t)}{\mathcal{I}_o(t)}.$$

We have

$$f'(t) + f(t)^2 \geq 0, \qquad f'_o(t) + f_o(t)^2 = 0. \tag{7.23}$$

Consider

$$h(t) := \Big\{f(t) - f_o(t)\Big\} \exp\Big\{ \int [f(t) + f_o(t)]dt\Big\}.$$

By (7.23), we obtain

$$h'(t) = \left\{ [f'(t) + f(t)^2] - [f_o'(t) + f_o(t)^2] \right\} \exp\left\{ \int [f(t) + f_o(t)]dt \right\} \geq 0.$$

This $h(t)$ is a non-decreasing function. Note that $h(0) = 0$. Thus $h(t) \leq 0$ for $t \leq 0$ and $h(t) \geq 0$ for $t \geq 0$. We obtain

$$\left(\ln\left| \frac{\mathcal{I}(t)}{\mathcal{I}_o(t)} \right| \right)' = f(t) - f_o(t) = \begin{cases} \leq 0 & \text{if } t \leq 0 \\ \geq 0 & \text{if } t \geq 0 \end{cases}.$$

We see that $\ln\left| \dfrac{\mathcal{I}(t)}{\mathcal{I}_o(t)} \right|$ attains its minimum at $t = 0$, that is,

$$\ln\left| \frac{\mathcal{I}(t)}{\mathcal{I}_o(t)} \right| \geq \ln\left| \frac{\mathcal{I}(0)}{\mathcal{I}_o(0)} \right| = 0.$$

Equivalently

$$|\mathcal{I}(t)| \geq |\mathcal{I}_o(t)|.$$

By a simple argument, one can see that the above inequality holds for t in the maximal interval (α, β) on which $\mathcal{I}_o(t) \neq 0$. This proves (7.22).

If $\mathcal{I}'(0)$ has the same sign as $\mathcal{I}(0)$, then

$$I(x, t) \geq \left| \mathcal{I}(t) \right| \geq \left| \mathcal{I}(0) \right| + \left| \mathcal{I}'(0) \right| t, \qquad t > 0.$$

Thus $\beta = +\infty$. Letting $t \to \infty$ yields that $\mathcal{I}(t)$ grows at least linearly. This is impossible. If $\mathcal{I}'(0)$ has the opposite sign as $\mathcal{I}(0)$, then

$$I(x, -t) \geq \left| \mathcal{I}(t) \right| \geq \left| \mathcal{I}(0) \right| - \left| \mathcal{I}'(0) \right| t, \qquad t < 0.$$

Thus $\alpha = -\infty$. Letting $t \to -\infty$ yields that $\mathcal{I}(t)$ grows at least linearly. This is impossible. Therefore, $\mathcal{I}'(0) = 0$, that is

$$\mathbf{J}_y(u) = 0.$$

This is a contradiction.

We have shown that $\mathbf{J} = 0$. Equation (7.5) is reduced to

$$\mathbf{K}(x)I_k = 0. \tag{7.24}$$

From (7.24), one can see that $I_k = 0$ at x where $\mathbf{K}(x) < 0$. Hence F_x is Euclidean by Deicke's theorem.

Now assume that $\mathbf{K} \le 0$ and the Cartan torsion grows sub-linearly. By (7.4),

$$C_{ijk|p|q}y^p y^q + \mathbf{K}F^2 C_{ijk} = 0. \tag{7.25}$$

By a similar argument, one can show that $L_{ijk} = C_{ijk|l}y^l = 0$. Thus F is Landsbergian. Q.E.D.

Let (M, F) be a closed Finsler manifold. Then F is always complete and the Cartan torsion is bounded. Further, assume that F has isotropic flag curvature $\mathbf{K} = \mathbf{K}(x)$. By Theorem 7.2.4, F is Riemannian when $\mathbf{K} < 0$ and F is Landsbergian when $\mathbf{K} = 0$. In the case when $\mathbf{K} = 0$, one can show that F is actually Berwaldian. Hence it is locally Minkowskian. We state this corollary without further details. See [86].

Corollary 7.2.5 ([1]) *Let (M, F) be a closed Finsler manifold with flag curvature $\mathbf{K} = \mathbf{K}(x)$. Then*

(a) if $\mathbf{K} < 0$, then F is Riemannian.
(b) if $\mathbf{K} = 0$, then it is locally Minkowskian.

We know that every Berwald metric is a Landsberg metric $(\mathbf{L} = 0)$. Thus we may ask the following question again: is there any non-Berwaldian Finsler metric satisfying

$$\mathbf{K} = 0, \qquad \mathbf{L} = 0 \ \ (\text{or } \mathbf{J} = 0)?$$

If such a metric exists, then by the above theorem, either it is incomplete or it has unbounded (mean) Cartan torsion.

7.3 Randers Metrics of Scalar Flag Curvature

It is one of the important problems in Finsler geometry to classify Finsler metrics of constant/scalar flag curvature. However, this problem still remains open even in the constant flag curvature case We shall first study Randers metrics of scalar flag curvature with isotropic S-curvature.

Let F be a Randers metric expressed in terms of a Riemannian metric $h = \sqrt{h_{ij}y^i y^j}$ and a vector field $V = V^i \frac{\partial}{\partial x^i}$ by (5.12), i.e.,

$$F = \frac{\sqrt{\lambda h^2 + V_0^2}}{\lambda} - \frac{V_0}{\lambda}, \qquad V_0 := V_i y^i, \tag{7.26}$$

where $\lambda := 1 - h(x, V)^2$. Let $\mathcal{R}_{ij} := \frac{1}{2}(V_{i|j} + V_{j|i})$, $\mathcal{R}_j := V^i \mathcal{R}_{ij}$, $\mathcal{R} := \mathcal{R}_j V^j$, $\mathcal{S}_{ij} := \frac{1}{2}(V_{i|j} - V_{j|i})$ and $\mathcal{S}_j := V^i \mathcal{S}_{ij}$. By Lemma 3.1.3, the spray coefficients G^i of F can be expressed in terms of the spray coefficients G^i_h of h and the covariant derivatives of V with respect to h as follow:

$$G^i = G^i_h - F\mathcal{S}^i{}_0 - \frac{1}{2}F^2(\mathcal{R}^i + \mathcal{S}^i) + \frac{1}{2}\left\{\frac{y^i}{F} - V^i\right\}\left\{2F\mathcal{R}_0 - \mathcal{R}_{00} - F^2\mathcal{R}\right\}. \quad (7.27)$$

By (7.27), one can express the Riemann curvature $R^i{}_k$ of F in terms of the Riemann curvature $\bar{R}^i{}_k$ of h and the covariant derivatives of V. Recall the formula (2.49):

$$R^i{}_k = 2\frac{\partial G^i}{\partial x^k} - \frac{\partial^2 G^i}{\partial x^m \partial y^k}y^m + 2G^m\frac{\partial^2 G^i}{\partial y^m \partial y^k} - \frac{\partial G^i}{\partial y^m}\frac{\partial G^m}{\partial y^k}. \quad (7.28)$$

Let

$$G^i = G^i_h + Q^i,$$

where

$$Q^i := -F\mathcal{S}^i{}_0 - \frac{1}{2}F^2(\mathcal{R}^i + \mathcal{S}^i) + \frac{1}{2}\left\{\frac{y^i}{F} - V^i\right\}\left\{2F\mathcal{R}_0 - \mathcal{R}_{00} - F^2\mathcal{R}\right\}.$$

Then

$$R^i{}_k = \bar{R}^i{}_k + 2Q^i{}_{|k} - [Q^i{}_{|m}]_{y^k}y^m + 2Q^m[Q^i]_{y^m y^k} - [Q^i]_{y^m}[Q^m]_{y^k}. \quad (7.29)$$

Here "|" denotes the horizontal covariant differentiation with respect to h.

From now on, we assume that F is of isotropic S-curvature, i.e., $\mathbf{S} = (n + 1)cF$ for some scalar function $c = c(x)$. By Proposition 5.2.5,

$$\mathcal{R}_{00} = -2ch^2. \quad (7.30)$$

Then the spray coefficients G^i are reduced to the following expression:

$$G^i = G^i_h + Q^i. \quad (7.31)$$

where

$$Q^i := -F\mathcal{S}^i{}_0 - \frac{1}{2}F^2\mathcal{S}^i + cFy^i.$$

By (5.20), we have

$$V_{i|j|k} = 2\left(c_{x^i}h_{jk} - c_{x^j}h_{ik} - c_{x^k}h_{ij}\right) - \bar{R}_{kpij}V^p.$$

Then

$$S^i{}_{k|0} = 2(h^{im}c_{x^m}y_k - c_{x^k}y^i) - \bar{R}_k{}^i{}_{mq}V^my^q,$$

$$S^i{}_{0|k} = 2(h^{im}c_{x^m}y_k - c_{x^m}y^m\delta^i_k) + \bar{R}_p{}^i{}_{kq}y^pV^q,$$

$$S^i{}_{|k} = 2cS^i{}_k - S^i{}_mS^m_k + 2(c_{x^m}V^m\delta^i_k - h^{im}c_{x^m}V_k) - \bar{R}_p{}^i{}_{kq}V^pV^q,$$

$$S^i{}_{|0} = 2cS^i{}_0 - S^i{}_mS^m_0 + 2(c_{x^m}V^my^i - h^{im}c_{x^m}V_0) - \bar{R}_p{}^i{}_{mq}V^pV^qy^m,$$

$$S^i{}_{0|0} = 2(h^{im}c_{|m}h^2 - c_{|0}y^i) - \bar{R}_p{}^i{}_{mq}y^py^qV^m,$$

where $\bar{R}_p{}^i{}_{kq}$ denotes the Riemann curvature tensor of h such that $\bar{R}_p{}^i{}_k :=$ $\bar{R}_p{}^i{}_{kq}y^py^q$. Further, we need the following formulas

$$F_{|k} = \frac{2cF(y_k - FV_k) + F(FS_k + S_{k0})}{A},$$

$$F_{|0} = 2cF^2 + \frac{F^2}{A}S_0,$$

$$(F_{y^m})_{|0} = \left(\frac{h^2}{A^3}S_0 + 2c\frac{F}{A}\right)\{y_k - FV_k\} - \frac{F^2}{A^2}S_0V_k - \frac{F}{A}S_{k0}.$$

where $A := \sqrt{\lambda h^2 + V_0^2}$. By (??) and the above identities, we first obtain the following very simple formula:

$$
\begin{aligned}
R^i{}_k = &\ \bar{R}_p{}^i{}_{kq}y^py^q - F\bar{R}_p{}^i{}_{kq}V^py^q - F\bar{R}_p{}^i{}_{kq}y^pV^q \\
&+ F^2\bar{R}_p{}^i{}_{kq}V^pV^q - F_{y^k}\bar{R}_p{}^i{}_{mq}y^py^qV^m + FF_{y^k}\bar{R}_p{}^i{}_{mq}y^qV^pV^m \\
&+ \left(\frac{3c_{x^m}y^m}{F} - c^2 - 2c_{x^m}V^m\right)\{F^2\delta^i_k - FF_{y^k}y^i\}. \qquad (7.32)
\end{aligned}
$$

It is surprised that all the terms with S^i or $S^i{}_k$ do not occur in (7.32). We can rewrite (7.32) as follows

$$
\begin{aligned}
R^i{}_k = &\ \bar{R}_p{}^i{}_{kq}(y^p - FV^p)(y^q - FV^q) \\
&- F_{y^k}\bar{R}_p{}^i{}_{mq}(y^p - FV^p)(y^q - FV^q)V^m \\
&+ \left(\frac{3c_{x^m}y^m}{F} - c^2 - 2c_{x^m}V^m\right)\{F^2\delta^i_k - FF_{y^k}y^i\}. \qquad (7.33)
\end{aligned}
$$

Let

$$\xi^i := y^i - F(x,y)V^i, \qquad \xi_k := h_{ik}\xi^i.$$

Let $\tilde{h} := h(x,\xi) = \sqrt{h_{pq}\xi^p\xi^q} = \sqrt{\xi_k\xi^k}$ and $\tilde{V}_0 := V_i\xi^i$. By (5.16), we have

$\tilde{h} = F$. Thus

$$y^i = \xi^i + \tilde{h}V^i.$$

Observe that

$$\begin{aligned}\lambda\tilde{h} = \lambda F &= \sqrt{\lambda h^2 + V_0} - V_0 \\ &= \sqrt{\lambda h^2 + V_0} - V_i(\xi^i + \tilde{h}V^i) \\ &= \sqrt{\lambda h^2 + V_0} - \tilde{V}_0 - \tilde{h}(1 - \lambda).\end{aligned}$$

This gives

$$\sqrt{\lambda h^2 + V_0} = \tilde{h} + \tilde{V}_0.$$

By the above identities, we obtain

$$F_{y^k} = \frac{\xi_k}{\tilde{h} + \tilde{V}_0},$$

$$F^2\delta^i_k - FF_{y^k}y^i = \tilde{h}^2\delta^i_k - \xi_k\xi^i - \frac{1}{\tilde{h} + \tilde{V}_0}\xi_k(\tilde{h}^2\delta^i_p - \xi_p\xi^i)V^p.$$

Let

$$\tilde{R}^i{}_k := \bar{R}_p{}^i{}_{kq}\xi^p\xi^q.$$

By (7.33), we obtain the following

Lemma 7.3.1   ([27]) *Let $F = \alpha + \beta$ be a Randers metric expressed by (7.26). Suppose that it has isotropic S-curvature, $\mathbf{S} = (n+1)cF$. Then for any scalar function $\mu = \mu(x)$ on M,*

$$R^i{}_k - \left(\frac{3c_{x^m}y^m}{F} + \mu - c^2 - 2c_{x^m}V^m\right)\left\{F^2\delta^i_k - FF_{y^k}y^i\right\}$$

$$= \tilde{R}^i{}_k - \mu\left(\tilde{h}^2\delta^i_k - \xi_k\xi^i\right) - \frac{\xi_k}{\tilde{h} + \tilde{V}_0}\left\{\tilde{R}^i{}_p - \mu\left(\tilde{h}^2\delta^i_p - \xi_p\xi^i\right)\right\}V^p (7.34)$$

By Lemma 7.3.1, we immediately obtain the following

Theorem 7.3.2   ([27]) *Let F be a Randers metric on n-dimensional manifold M defined by (7.26). Suppose that $\mathbf{S} = (n+1)cF$ where $c = c(x)$ is a scalar function. Then F is of scalar flag curvature if and only if h is of* *sectional curvature* $\bar{\mathbf{K}} = \mu$, *where* $\mu = \mu(x)$ *is a scalar function* (=*constant if* $n \geq 3$). *In this case, the flag curvature of* F *is given by*

$$\mathbf{K} = \frac{3c_{x^m}y^m}{F} + \sigma, \qquad (7.35)$$

where $\sigma := \mu - c^2 - 2c_{x^m}V^m$.

Proof: Assume that F is of scalar flag curvature, then Proposition 7.1.2, the flag curvature is given by

$$\mathbf{K} = \frac{3c_{x^m}y^m}{F} + \sigma,$$

where $\sigma = \sigma(x)$ is a scalar function on M. Let

$$\mu := \sigma + c^2 + 2c_{x^m}V^m.$$

It suffices to show that h has sectional curvature $\bar{\mathbf{K}} = \mu$. It follows from (7.34) that

$$\tilde{R}^i_{\ k} - \mu\left(\tilde{h}^2\delta^i_k - \xi_k\xi^i\right) - \frac{1}{\tilde{h} + \tilde{V}_0}\xi_k\left\{\tilde{R}^i_{\ p} - \mu\left(\tilde{h}^2\delta^i_p - \xi_p\xi^i\right)\right\}V^p = 0.$$

Clearly, we have

$$\tilde{R}^i_{\ k} = \mu\left(\tilde{h}^2\delta^i_k - \xi_k\xi^i\right). \qquad (7.36)$$

Thus h has sectional curvature $\bar{K} = \mu$. By the Schur lemma (Lemma 7.1.1), $\mu = constant$ in dimension $n \geq 3$.

   Conversely, if h has sectional curvature $\bar{K} = \mu$, then (7.36) holds. By (7.34) again, we get

$$R^i_{\ k} = \left(\frac{3c_{x^m}y^m}{F} + \sigma\right)\left\{F^2\delta^i_k - FF_{y^k}y^i\right\}, \qquad (7.37)$$

where $\sigma = \mu - c^2 - 2c_{x^m}V^m$. Thus F is of scalar flag curvature.   Q.E.D.

   By Proposition 5.2.10 and Theorem 7.3.2, we obtain the following

Theorem 7.3.3   *Let* $F = \alpha + \beta$ *be a Randers metric on a manifold* M *of dimension* $n \geq 3$, *which is expressed in terms of a Riemannian metric* h *and a vector field* V *by* (7.26). F *is of isotropic S-curvature* $\mathbf{S} = (n+1)cF$

and of scalar flag curvature, $\mathbf{K} = K(x, y)$, *if and only if at any point, there is a local coordinate system in which h, c and V are given by*

$$h = \frac{\sqrt{|y|^2 + \mu(|x|^2|y|^2 - \langle x, y \rangle^2)}}{1 + \mu|x|^2}, \tag{7.38}$$

$$c = \frac{\delta + \langle a, x \rangle}{\sqrt{1 + \mu|x|^2}}, \tag{7.39}$$

$$V = -2\Big\{ \Big(\delta\sqrt{1 + \mu|x|^2} + \langle a, x \rangle \Big) x - \frac{|x|^2 a}{\sqrt{1 + \mu|x|^2} + 1} \Big\}$$
$$+ xQ + b + \mu\langle b, x \rangle x, \tag{7.40}$$

where δ, μ *are constants,* $Q = (q_j{}^i)$ *is an anti-symmetric matrix and* $a, b \in \mathbf{R}^n$ *are constant vectors. In this case, the flag curvature is given by*

$$\mathbf{K} = \frac{3c_{x^m} y^m}{F} + \sigma, \tag{7.41}$$

where $\sigma = \mu - c^2 - 2c_{x^m} V^m$.

Proof: By assumption, the dimension of M is not less than 3. First we assume that $F = \alpha + \beta$ is of isotropic S-curvature and of scalar flag curvature. By Theorem 7.3.2, the flag curvature of F is given by (7.35) and h has constant sectional curvature $\bar{\mathbf{K}} = \mu$. At any point, there is a local coordinate system in which h is given by (7.38). By Proposition 5.2.10, if $\mathbf{S} = (n+1)cF$, then c and V are given by (7.39) and (7.40) respectively in the same local coordinate system.

Conversely, assume that there is a local coordinate system in which h, c and V are given by (7.38), (7.39) and (7.40) respectively, then by Proposition 5.2.10, $\mathbf{S} = (n+1)cF$. Since h has constant sectional curvature $\bar{\mathbf{K}} = \mu$, by Theorem 7.3.2, F is of scalar curvature with flag curvature given by (7.35). Q.E.D.

Let us take a look at the following special example.

Example 7.3.4 ([93]) In (7.38)-(7.40), letting $\mu = 0, \delta = 0, Q = 0$ and $b = 0$, we get

$$h = |y|, \quad c = \langle a, x \rangle, \quad V = -2\langle a, x \rangle x + |x|^2 a. \tag{7.42}$$

The Randers metric $F = \alpha + \beta$ defined by (7.26) is given by

$$F = \frac{\sqrt{(1 - |a|^2|x|^4)|y|^2 + (|x|^2\langle a, y\rangle - 2\langle a, x\rangle\langle x, y\rangle)^2}}{1 - |a|^2|x|^4}$$
$$- \frac{|x|^2\langle a, y\rangle - 2\langle a, x\rangle\langle x, y\rangle}{1 - |a|^2|x|^4}.$$

The above defined Randers metric F is of isotropic S-curvature and scalar flag curvature, i.e.,

$$\mathbf{S} = (n+1)\langle a, x\rangle\, F, \qquad \mathbf{K} = \frac{3\langle a, x\rangle}{F} + 3\langle a, x\rangle^2 - 2|a|^2|x|^2.$$

This example satisfies the assumption and conclusion of Proposition 7.1.2.

Now we study the Ricci curvature of a Randers metric with isotropic S-curvature. Let $\mathbf{Ric}$ and $\overline{\mathbf{Ric}}$ denote the Ricci curvature of F and h respectively. They are defined by

$$\mathbf{Ric} := R^m_{\ m}, \qquad \overline{\mathbf{Ric}} := \bar{R}^m_{\ m}.$$

Let

$$\widetilde{\mathbf{Ric}} := \tilde{R}^m_{\ m} = \bar{R}_p{}^m{}_{mq}\xi^p\xi^q,$$

where $\xi^i := y^i - FW^i$. Clearly, $\overline{\mathbf{Ric}} = (n-1)\mu h^2$ if and only if $\widetilde{\mathbf{Ric}} = (n-1)\mu\tilde{h}^2$.

First we have the following

Lemma 7.3.5 ([27]) *Let $F = \alpha + \beta$ be a Randers metric expressed by (7.26). Suppose that it has isotropic S-curvature, $\mathbf{S} = (n+1)cF$. Then for any scalar function $\mu = \mu(x)$ on M,*

$$\mathbf{Ric} - (n-1)\left(\frac{3c_{x^m}y^m}{F} + \mu - c^2 - 2c_{x^m}V^m\right)F^2 = \widetilde{\mathbf{Ric}} - (n-1)\mu\tilde{h}^2. \quad (7.43)$$

Proof: Observe that

$$\xi_m\tilde{R}^m_p = \xi_m\tilde{R}_i{}^m{}_{pj}\xi^i\xi^j = \xi^m\tilde{R}_{impj}\xi^i\xi^j = 0$$

and

$$\xi_m\left(\tilde{h}^2\delta^m_p - \xi_p\xi^m\right) = \tilde{h}^2\xi_p - \xi_p\tilde{h}^2 = 0.$$

Then (7.43) follows from (7.34). Q.E.D.

From Lemma 7.3.5 we immediately obtain the following

Theorem 7.3.6 ([27]) *Let F be a Randers metric on n-dimensional manifold M defined by (7.26) and let $c = c(x)$ and $\mu = \mu(x)$ be scalar functions on M. Suppose $\mathbf{S} = (n+1)cF$. Then $\overline{\mathbf{Ric}} = (n-1)\mu h^2$ if and only if*

$$\mathbf{Ric} = (n-1)\left\{\frac{3c_{x^m}y^m}{F} + \mu - c^2 - 2c_{x^m}V^m\right\}F^2. \qquad (7.44)$$

In general, a Randers metric of scalar flag curvature does not necessarily have isotropic S-curvature. However, we have the following

Lemma 7.3.7 ([9]) *Let F be a Randers metric on n-dimensional manifold M defined by (7.26) using a Riemannian metric h and a vector field V. If the Ricci curvature is constant, $\mathbf{Ric} = (n-1)\kappa F^2$ where κ is a constant, then there is a constant c such that*

$$\mathcal{R}_{00} = -2ch^2.$$

The proof of Lemma 7.3.7 is technical, so is omitted.

Now we can give a complete list of explicit formulas for Randers metrics of constant flag curvature. Let F be a Randers metric defined by (7.26). Suppose that F has constant flag curvature. Then by Lemma 7.3.7, V satisfies

$$\mathcal{R}_{00} = -2ch^2, \qquad c = constant.$$

That is, F has constant S-curvature. By Theorem 7.3.2, h has sectional curvature $\bar{\mathbf{K}} = \mu$ where $\mu = \mu(x)$ is a scalar function. In this case, the flag curvature of F is given by

$$\mathbf{K} = \mu - c^2.$$

Since $\mathbf{K} = constant$ by assumption, we conclude that $\mu = constant$. Namely, h has constant sectional curvature $\bar{\mathbf{K}} = \mu$. By chooing a local coordinate system, we may assume that $h = h_\mu$ is defined in (7.38),

$$h_\mu = \frac{\sqrt{|y|^2 + \mu(|x|^2|y|^2 - \langle x, y\rangle^2)}}{1 + \mu|x|^2}. \qquad (7.45)$$

By Corollary 5.2.11, V is given by

$$V = \begin{cases} -2cx + xQ + b & \text{if } \mu = 0 \\ xQ + b + \mu\langle b, x\rangle x & \text{if } \mu \neq 0, \end{cases} \tag{7.46}$$

where $Q = (q_j{}^i)$ is a skew-symmetric matrix and $b = (b^i)$ is a constant vector with $|b| < 1$. This completes the proof of the following

Theorem 7.3.8 ([7]) *Let F be a Randers metric on a manifold M, whichis expressed by (7.26) in terms of a Riemannian metric h and a vector field V. Then F has constant flag curvature if and only if h is a Riemannian metric and V is a vector field on M with the following property: at any point in M, there is a local coordinate system (x^i) with $x^i(p) = 0$ such that h is locally expressed by (7.45) and $V = (V^i)$ is given by (7.46). In this case, $\mathbf{S} = (n + 1)cF$ and $\mathbf{K} = \mu - c^2$.*

According to Theorem 7.3.8, any Randers metric with $\mathbf{K} = 0$ is given by

$$F = \frac{\sqrt{|y|^2 - (|y|^2|xQ + b|^2) - \langle xQ + b, \, y\rangle^2)}}{1 - |xQ + b|^2} - \frac{\langle xQ + b, y\rangle}{1 - |xQ + b|^2}.$$

One can show that F is positively complete if and only if $Q = 0$, in which case, F is Minkowskian. This fact can also be proved in a different way without using the above classification result (see Theorem 1.2 in [90]).

By Theorem 7.3.8, one can show that there are non-Riemannian Randers metrics of constant flag curvature $\mathbf{K} = 1$ on any standard unit sphere $S^n \subset R^{n+1}$, regardless of the dimension. See [7] and [13] for discussions on special Randers metrics of $\mathbf{K} = 1$ on S^n. The geodesic properties of some examples in Theorem 7.3.8 have been discussed in [50] and [103]. See [82] for more recent work on geodesics.

The classification problem of Randers metrics with constant flag curvature was first attempted by Yasuda-Shimada [101] and Matsumoto [67]. Motivated by Yasuda-Shimada's result, Bao-Shen constructed a family of Randers metrics on S^3 with $\mathbf{K} = 1$ [10]. Bao-Shen's examples satisfy the conditions listed in [101] and [67] for Randers metrics of positive constant flag curvature. It was believed that Yasuda-Shimada's result was completely true, until some new counter-examples were constructed [90] and

[91]. Shortly after these examples were found, a correct version of Yasuda-Shimada's result was obtained by Bao-Robles [8], meanwhile, independently by Matsumoto-Shimada [71]. From their results, one learns that Yasuda-Shimada's conclusions are still true under an additional condition. This additional condition is satisfied by the example in [10] (see also [12] [13]).

We have seen many irreversible Finsler metrics of constant flag curvature. We know that the Hilbert metric $H = H(x, y)$ on a strongly convex domain $\mathcal{U}$ is reversible complete with $\mathbf{K} = -1$. Is there any reversible Finsler metrics with positive constant flag curvature on S^n? In [84], the following theorem is proved: if (M, F) is an n-dimensional closed simply connected reversible Finsler manifold with $\mathbf{K} = 1$, then M is diffeomorphic to S^n and for any point $p \in M$ there is a unique point p^* with $d(p, p^*) = \pi$ such that any unit speed geodesic from p must pass p^* and form a closed geodesic of length 2π. In [53], Kim-Yim prove that any reversible Finsler metric on S^n with $\mathbf{K} = 1$ and $\mathbf{S} = 0$ must be Riemannian. The proof requires a volume comparison theorem on the metric balls in a Finsler manifold [85] and a volume comparison theorem on the unit sphere bundle of a Finsler manifold [35]. Recently, Foulon claims that any reversible Finsler metric on S^2 with $\mathbf{K} = 1$ is Riemannian, provided that its geodesics are great circles [38]. Most recently, R. Bryant announces that any reversible Finsler metric on S^2 with $\mathbf{K} = 1$ must be Riemannian. Bryant's proof is a combination of his earlier results [20] with a fundamental (and deeper) result of LeBrun and Mason [63].

Chapter 8

Projectively Flat Finsler Metrics

It is Hilbert's Fourth Problem to characterize the (not-necessarily-reversible) distance functions on an open subset in R^n such that straight lines are geodesics [44]. Regular distance functions with straight geodesics are projectively flat Finsler metrics. They are characterized by a system of partial differential equations (3.18). However, it is still difficult to understand the local metric structure of such metrics. In this chapter, we will discuss projectively flat metrics with special curvature properties.

8.1 Projectively Flat Randers Metrics

According to Proposition 3.4.8, a Randers metric $F = \alpha + \beta$ is locally projectively flat if and only if α is locally projectively flat and β is closed. By the Beltrami theorem in Riemannian geometry, α is locally projectively flat if and only if it is of constant sectional curvature $\mathbf{K}_\alpha = \mu$. In this case, α is locally isometric to the following metric defined on a ball $B^n(r_\mu) \subset R^n$.

$$\alpha_\mu := \frac{\sqrt{|y|^2 + \mu(|x|^2|y|^2 - \langle x, y \rangle^2)}}{1 + \mu|x|^2}. \tag{8.1}$$

Moreover, if F is of constant flag curvature or isotropic S-curvature, β can be completely determined.

First, let us consider projectively flat Randers metrics with constant flag curvature.

Proposition 8.1.1 ([88]) *Let $F = \alpha + \beta$ ($\beta \not\equiv 0$) be a locally projectively flat Randers metric on an n-dimensional manifold M. Suppose that it has*

149

constant Ricci curvature $\mathbf{Ric} = (n-1)\lambda F^2$. *Then* $\lambda \leq 0$. *If* $\lambda = 0$, F *is locally Minkowskian. If* $\lambda = -1/4$, F *is given by*

$$F = \frac{\sqrt{|y|^2 - (|x|^2|y|^2 - \langle x, y\rangle^2)} \pm \langle x, y\rangle}{1 - |x|^2} \pm \frac{\langle a, y\rangle}{1 + \langle a, x\rangle}, \qquad (8.2)$$

where $a \in \mathbf{R}^n$ *is a constant vector. In this case,*

$$\mathbf{K} = -\frac{1}{4}, \qquad \mathbf{S} = \pm\frac{1}{2}(n+1)F.$$

Proof: One may assume that α has constant sectional curvature $\mathbf{K}_\alpha = \mu$ and β is closed. Let $\Phi = b_{i;j}y^i y^j$ and $\Psi = b_{i;j;k}y^i y^j y^k$ be the homogeneous functions defined in (6.11). Since β is closed, $\Phi = e_{00}$.

It follows from (6.10) that

$$\mu\alpha^2 + 3\left(\frac{\Phi}{2F}\right)^2 - \frac{\Psi}{2F} = \lambda F^2.$$

That is,

$$\mu\alpha^2(\alpha + \beta)^2 + \frac{3}{4}\Phi^2 - \frac{1}{2}\Psi(\alpha + \beta) = \lambda(\alpha + \beta)^4.$$

This gives rise to two equations

$$\frac{3}{4}\Phi^2 = \frac{1}{2}\beta\Psi + (\lambda - \mu)\alpha^4 + (6\lambda - \mu)\alpha^2\beta^2 + \lambda\beta^4, \qquad (8.3)$$

$$\frac{1}{2}\Psi = (2\mu - 4\lambda)\alpha^2\beta - 4\lambda\beta^3. \qquad (8.4)$$

Substituting (8.4) into (8.3) yields

$$\frac{3}{4}\Phi^2 = (\lambda - \mu)\alpha^4 + (2\lambda + \mu)\alpha^2\beta^2 - 3\lambda\beta^4. \qquad (8.5)$$

Differentiating (8.5), one obtains

$$\frac{3}{2}\Phi\, b_{i;j;k}y^i y^j = 2(2\lambda + \mu)\alpha^2\beta\, b_{i;k}y^i - 12\lambda\beta^3\, b_{i;k}y^i. \qquad (8.6)$$

Contracting (8.6) with y^k yields

$$\frac{3}{2}\Phi\Psi = 2\left\{(2\lambda + \mu)\alpha^2 - 6\lambda\beta^2\right\}\Phi\beta. \qquad (8.7)$$

Substituting (8.4) into (8.7) yields

$$4(\mu - 4\lambda)\Phi\alpha^2\beta^2 = 0.$$

We assert that $\mu = 4\lambda$. If not, from the above equation, $\Phi\beta \equiv 0$. Then on the open subset $U := \{x \in M \mid \beta_x \neq 0\}$, $\Phi = 0$. Thus β is parallel with respect to α and $\Psi = 0$. It follows from (8.4) that $\mu = 4\lambda = 0$. It is a contradiction.

Substituting $\mu = 4\lambda$ into (8.5) yields

$$\Phi^2 = -4\lambda(\alpha^2 - \beta^2)^2.$$

It follows that $\lambda \leq 0$. Let $c := \pm\sqrt{-\lambda}$. The above identity becomes

$$e_{00} = 2c(\alpha^2 - \beta^2). \tag{8.8}$$

Now we are going to find an explicit formula for β using (8.8).

Assume that $\lambda = 0$. By the above argument, $\mu = 4\lambda = 0$ and $c = \pm\sqrt{-\lambda} = 0$. Thus α is flat and $e_{00} = 0$. Since β is closed, we get $b_{i;j} = 0$, namely, β is parallel with respect to α. In this case, $G^i(x,y) = G^i_\alpha(x,y)$ quadratic in y. F is a Berwald metric with zero flag curvature. Thus it is locally Minkowskian by Theorem 2.3.2.

Assume that $\lambda = -c^2 < 0$. In this case, α has negative constant curvature $\mu = -4c^2$. By scaling, we may assume that α has constant curvature $\mu = -1$ ($c = \pm\frac{1}{2}$), hence $\lambda = -1/4$. Thus α can be expressed as α_{-1},

$$\alpha_{-1} = \frac{\sqrt{|y|^2 - (|x|^2|y|^2 - \langle x,y\rangle^2)}}{1 - |x|^2}, \qquad y \in T_x B^n(1) \cong R^n.$$

It follows from (8.8) that

$$b_{i;j} = \varepsilon(a_{ij} - b_i b_j), \qquad \varepsilon = \pm 1. \tag{8.9}$$

We can express $\beta = b_i y^i$ in the following gradient form,

$$\beta = \varepsilon \frac{\langle x,y\rangle}{1 - |x|^2} + \varepsilon \frac{df_x(y)}{f(x)},$$

where $f(x) > 0$ is a scalar function on $B^n(1)$ and $\varepsilon = \pm 1$ is the same as in (8.9). It follows from (8.9) that $f_{x^i x^j} = 0$. Thus f is a linear function

$$f = \delta(1 + \langle a,x\rangle), \qquad \delta > 0,$$

and

$$\beta = \varepsilon \frac{\langle x,y\rangle}{1 - |x|^2} + \varepsilon \frac{\langle a,y\rangle}{1 + \langle a,x\rangle}, \qquad y \in T_x B^n(1) \cong R^n.$$

Q.E.D.

Next we are going to study projectively flat Randers metrics with almost isotropic S-curvature. Recall that for a locally projectively flat Randers metric $F = \alpha + \beta$, the Riemannian metric $\alpha \cong \alpha_\mu$ must be of constant sectional curvature $\mathbf{K}_\alpha = \mu$ and the 1-form β must be closed. We have the following

Proposition 8.1.2   ([24]) *Let $F = \alpha + \beta$ be a locally projectively flat Randers metric on an n-dimensional manifold M. Suppose that F has almost isotropic S-curvature*

$$\mathbf{S} = (n+1)\Big\{cF + \eta\Big\}, \tag{8.10}$$

where $c = c(x)$ is a scalar function on M and $\eta = \eta_i(x)y^i$ is a closed 1-form on M. Then the flag curvature is given by

$$\mathbf{K} = \frac{3c_{x^k}(x)y^k}{F(x,y)} + 3c(x)^2 + \mu \tag{8.11}$$

$$= \frac{3}{4}\Big\{\mu + 4c(x)^2\Big\}\frac{F(x,-y)}{F(x,y)} + \frac{\mu}{4}, \tag{8.12}$$

where c_{x^k} denotes the partial derivative of c with respect to x^k. Moreover,

(A) *if $\mu + 4c(x)^2 \equiv 0$, then $c(x) = c$ is a constant and the flag curvature $\mathbf{K} = -c^2$. In this case, $F = \alpha + \beta$ is either locally Minkowskian ($c = 0$) or, up to a scaling ($c = \pm 1/2$), locally isometric to the metric in (8.2);*

(B) *if $\mu + 4c(x)^2 \neq 0$, then $F = \alpha + \beta$ is locally given by*

$$\alpha \cong \alpha_\mu(x,y), \qquad \beta = -\frac{2c_{x^k}(x)y^k}{\mu + 4c(x)^2} \tag{8.13}$$

where α_μ is given by (8.1) and $c(x) := c_\mu(x)$ is given by

$$c_\mu(x) = \begin{cases} (\lambda + \langle a, x \rangle)\sqrt{\pm(1+\mu|x|^2)-(\lambda+\langle a,x\rangle)^2} & \text{if } \mu \neq 0 \\[2mm] \dfrac{\pm 1}{2\sqrt{\lambda + 2\langle a,x\rangle + |x|^2}} & \text{if } \mu = 0, \end{cases} \tag{8.14}$$

where $a \in R^n$ is a constant vector and λ is a constant number.

Proof: Let $\alpha = \sqrt{a_{ij}y^iy^j}$ and $\beta = b_iy^i$. Let $\Phi = b_{i;j}y^iy^j$ and $\Psi = b_{i;j;k}y^iy^jy^k$ be the homogeneous functions defined in (6.11). Since β is closed, $\Phi = e_{00}$. By Lemma 5.2.2,

$$\Phi = e_{00} = 2c(\alpha^2 - \beta^2) = 2c(\alpha - \beta)F. \tag{8.15}$$

Using (8.15), one obtains from $\alpha_{;k} = 0$, $\Phi = \beta_{;k}y^k$ and $\Psi = \Phi_{;k}y^k$ that

$$
\begin{aligned}
\Psi &= 2c_{;k}y^k(\alpha^2 - \beta^2) - 4c\beta\beta_{;k}y^k \\
&= 2c_{x^k}y^k(\alpha^2 - \beta^2) - 4c\Phi\beta \\
&= 2c_{x^k}y^k(\alpha^2 - \beta^2) - 8c^2\beta(\alpha^2 - \beta^2) \\
&= 2(c_{x^k}y^k - 4c^2\beta)(\alpha - \beta)F.
\end{aligned}
$$

By (6.10) and the above formulas, one obtains

$$
\begin{aligned}
\mathbf{K}F^2 &= \mu\alpha^2 + 3\left[\frac{\Phi}{2F}\right]^2 - \frac{\Psi}{2F} \\
&= \mu\alpha^2 + 3c^2(\alpha - \beta)^2 - (c_{x^k}y^k - 4c^2\beta)(\alpha - \beta).
\end{aligned} \qquad (8.16)
$$

On the other hand, by Theorem 7.1.2, the flag curvature is in the following form

$$
\mathbf{K} = \frac{3c_{x^k}y^k}{F} + \sigma, \qquad (8.17)
$$

where $\sigma = \sigma(x)$ is a scalar function on M. Combining (8.16) and (8.17) yields

$$
2\left\{2c_{x^k}y^k + (\sigma + c^2)\beta\right\}\alpha + \left\{2c_{x^k}y^k + (\sigma + c^2)\beta\right\}\beta \\
+ \left\{\sigma - 3c^2 - \mu\right\}\alpha^2 = 0.
$$

This gives

$$
2c_{x^k}y^k + (\sigma + c^2)\beta = 0, \qquad (8.18)
$$
$$
\sigma - 3c^2 - \mu = 0. \qquad (8.19)
$$

From (8.19), one obtains that $\sigma = 3c^2 + \mu$. Substituting it into (8.18) yields

$$
(\mu + 4c^2)\beta = -2c_{x^k}y^k. \qquad (8.20)
$$

Substituting $\sigma = 3c^2 + \mu$ into (8.17) yields (8.11). Using (8.20), one can rewrite (8.11) as (8.12).

Case 1: Suppose that $\mu + 4c(x)^2 \equiv 0$. Then $c(x) = constant$. It follows from (8.11) that

$$
\mathbf{K} = 3c^2 + \mu = -c^2.
$$

Then Theorem 8.1.2 (A) follows from Proposition 8.1.1.

Case 2: Suppose that $\mu + 4c^2 \neq 0$ on an open subset $\mathcal{U} \subset M$. Then by (8.20), β is given by

$$\beta = -\frac{2c_{x^k}y^k}{\mu + 4c^2}. \tag{8.21}$$

Note that β is exact. Let $c_{;i} := c_{x^i}$ and $c_{;i;j}$ denote the covariant derivatives of c with respect to α. It follows from (8.15) and (8.21) that

$$c_{;i;j} = -c(\mu + 4c^2)a_{ij} + \frac{12cc_{;i}c_{;j}}{\mu + 4c^2}.$$

The above equation can be converted to the following equation in a local coordinate system:

$$c_{x^i x^j} + \frac{\mu(x^i c_{x^j} + x^j c_{x^i})}{1 + \mu|x|^2} =$$

$$-c(\mu + 4c^2)\Big\{\frac{\delta_{ij}}{1 + \mu|x|^2} - \frac{\mu x^i x^j}{(1 + \mu|x|^2)^2}\Big\} + \frac{12cc_{x^i}c_{x^j}}{\mu + 4c^2}. \tag{8.22}$$

To solve (8.22) for $c = c(x)$, let

$$\varphi := \begin{cases} \frac{2c\sqrt{1+\mu|x|^2}}{\sqrt{\pm(\mu+4c^2)}} & \text{if } \mu \neq 0 \\ \frac{1}{c^2} & \text{if } \mu = 0 \end{cases} \tag{8.23}$$

where the sign $\pm$ depends on the value of c such that $\pm(\mu + 4c^2) > 0$. Then (8.22) is reduced to the following equation:

$$\varphi_{x^i x^j} = \begin{cases} 0 & \text{if } \mu \neq 0 \\ 8\delta_{ij} & \text{if } \mu = 0 \end{cases}.$$

One immediately obtains

$$\varphi = \begin{cases} \lambda + \langle a, x \rangle & \text{if } \mu \neq 0 \\ 4(\lambda + 2\langle a, x \rangle + |x|^2) & \text{if } \mu = 0 \end{cases},$$

where $a \in R^n$ is a constant vector and λ is a constant number. Solving (8.23) for c gives the formula in (8.14).   Q.E.D.

By Proposition 8.1.2, one immediately obtains the following

Corollary 8.1.3 *Let $F = \alpha + \beta$ be a locally projectively flat Randers metric on an n-dimensional manifold M. Suppose that F has constant S-curvature $\mathbf{S} = (n+1)cF$. Then F is locally Minkowskian, or Riemannian with constant curvature, or up to a scaling, locally isometric to the metric in (8.2).*

Proof: Let μ be the constant sectional curvature of α. First assume that $\mu + 4c^2 = 0$. Then by Proposition 8.1.2 (A), $F = \alpha + \beta$ is either locally Minkowskian or, up to a scaling, locally isometric to the metric in (8.2). Suppose that $\mu + 4c^2 \neq 0$. Then by Proposition 8.1.2 (B), $F = \alpha + \beta$ is given by (8.13). Since $c_{x^k} = 0$, $\beta = 0$ and $F = \alpha$ is a Riemannian metric.

Q.E.D.

8.2 Projectively Flat Metrics with Constant Flag Curvature

The first set of Finsler metrics of constant flag curvature were discovered by Hilbert, Berwald [17] and Funk [39], [40]. All these metrics are locally projectively flat. In the past ten years, R. Bryant made a significant progress in the study of Finsler metrics of constant flag curvature. In particularly, he has classified projectively flat Finsler metrics on S^n with constant curvature $\mathbf{K} = 1$ [19], [20], [21]. In this section, we shall study and characterize projectively flat Finsler metrics of constant flag curvature. Such metrics can be described using algebraic equations or using Taylor expansions. We shall also discuss various examples.

Lemma 8.2.1 *Let $F = F(x, y)$ be a Finsler metric on an open subset $\mathcal{U} \subset \mathbf{R}^n$. Then F is projectively flat with constant flag curvature $\mathbf{K} = \lambda$ if and only if there are positively y-homogeneous functions $P = P(x, y)$ on $T\mathcal{U} \cong \mathcal{U} \times \mathbf{R}^n$ such that*

$$F_{x^k} = (PF)_{y^k}, \tag{8.24}$$

$$P_{x^k} = PP_{y^k} - \lambda FF_{y^k}, \tag{8.25}$$

in which case, $P = \frac{1}{2F}F_{x^m}y^m$ is the projective factor of F and $P^2 - P_{x^k}y^k = \lambda F^2$.

Proof: Assume that $F = F(x, y)$ is projectively flat on $\mathcal{U}$. Then it satisfies (3.18) and the projective factor is given by $P := \frac{1}{2F} F_{x^m} y^m$. Observe that

$$(PF)_{y^k} = \frac{1}{2}(F_{x^m} y^m)_{y^k} = \frac{1}{2}(F_{x^m y^k} y^m + F_{x^k}) = \frac{1}{2}(F_{x^k} + F_{x^k}) = F_{x^k}.$$

Thus F satisfies (8.24). Comparing (7.29) and (6.7) yields

$$P_{x^k} - PP_{y^k} = -\frac{(\mathbf{K}F^3)_{y^k}}{3F} = -\lambda F F_{y^k}.$$

Thus $P = P(x, y)$ satisfies (8.25).

Conversely, suppose that (8.24) and (8.25) hold for some positively y-homogeneous functions $P = P(x, y)$ and a constant λ. First by (8.24), one obtains

$$F_{x^k y^l} y^k = (PF)_{y^k y^l} y^k = (PF)_{y^l} = F_{x^l},$$

$$\frac{1}{2F} F_{x^k} y^k = \frac{1}{2F}(PF)_{y^k} y^k = P.$$

By Theorem 3.3.1, F is projectively flat and P is the projective factor. Contracting (8.25) with y^k yields that

$$P^2 - P_{x^k} y^k = \lambda F^2.$$

By (6.8), the flag curvature is a constant, i.e., $\mathbf{K} = \lambda$. Q.E.D.

Before we go on, let us take a look at the Funk metric.

Example 8.2.2 Let $\phi = \phi(y)$ be a Minkowski norm on $\mathbb{R}^n$ and $\mathcal{U}$ be the domain enclosed by the indicatrix of ϕ. The Funk metric $\Theta = \Theta(x, y)$ on $\mathcal{U}$ is defined by

$$\Theta(x, y) = \phi\Big(y + \Theta(x, y)\, x\Big), \qquad y \in T_x\mathcal{U}. \tag{8.26}$$

It satisfies (1.38), i.e., $\Theta_{x^k} = \Theta\Theta_{y^k}$. By this PDE, one can easily verify that $F := \Theta(x, y)$ and $P := \frac{1}{2}\Theta(x, y)$ satisfy (8.24) and (8.25) with $\lambda = -1/4$, respectively. Thus Θ is projectively flat with constant flag curvature $\mathbf{K} = -1/4$. When $\mathcal{U} = B^n(1)$ is the standard unit ball in $\mathbb{R}^n$, Θ is given by

$$\Theta = \frac{\sqrt{(1 - |x|^2)|y|^2 + \langle x, y \rangle^2}}{1 - |x|^2} + \frac{\langle x, y \rangle}{1 - |x|^2}. \tag{8.27}$$

We can construct projectively flat Finsler metrics of negative constant flag curvature using algebraic equations.

Theorem 8.2.3 ([89]) *Let $\psi = \psi(y)$ be an arbitrary Minkowski norm on R^n and $\varphi = \varphi(y)$ be an arbitrary positively homogeneous function of degree one on R^n. For x close to the origin and $y \in \mathrm{R}^n$, define $\Psi_\pm = \Psi_\pm(x, y)$ by*

$$\Psi_\pm(x, y) = \phi_\pm\Big(y + \Psi_\pm(x, y)\, x\Big), \tag{8.28}$$

where $\phi_\pm := \varphi(y) \pm \psi(y)$. Then

$$F := \frac{1}{2}\Big\{\Psi_+(x, y) - \Psi_-(x, y)\Big\} \tag{8.29}$$

is a projectively flat Finsler metric with constant flag curvature $\mathbf{K} = -1$ and $F(0, y) = \psi(y)$, and its projective factor $P = P(x, y)$ is given by

$$P = \frac{1}{2}\Big\{\Psi_+(x, y) + \Psi_-(x, y)\Big\} \tag{8.30}$$

with $P(0, y) = \varphi(y)$.

We first prove that the functions $\Psi_\pm$ defined in (8.28) exist and satisfy the following equations:

$$(\Psi_\pm)_{x^k} = \Psi_\pm (\Psi_\pm)_{y^k}.$$

Then, using these equations, we can show that the functions F and P defined in (8.29) and (8.30) have the desired properties.

Lemma 8.2.4 *Let $\phi = \phi(y)$ be an arbitrary positively homogeneous function of degree one on R^n. Suppose that ϕ is C^∞ on $\mathrm{R}^n \setminus \{0\}$. Then for x close to the origin, there is a unique real-valued function $f := f(x, y)$ satisfying the following*

$$f(x, y) = \phi\Big(y + f(x, y)\, x\Big). \tag{8.31}$$

Moreover, f satisfies

$$f_{x^k} = f f_{y^k}. \tag{8.32}$$

Proof. Fix $y \in T_x R^n \cong R^n$ with $y \neq 0$. Let $h(t) := t - \phi(y + tx)$. By the homogeneity of ϕ, there is a small $\delta > 0$ such that if $|x| < \delta$, then at any t with $y + tx \neq 0$,

$$h'(t) = 1 - \phi_{y^k}(y + tx)x^k \geq \frac{1}{2}.$$

Note that $h(0) < 0$. Thus there is a unique $t_o > 0$ such that $h(t_o) = 0$. Setting $f(x, y) := t_o$, one obtains the unique solution.

Differentiating (8.31) with respect to x^k and y^k gives

$$\left(1 - \phi_{y^m} x^m\right) f_{x^k} = \phi_{y^k} f$$

$$\left(1 - \phi_{y^m} x^m\right) f_{y^k} = \phi_{y^k}.$$

Since $\phi_{y^m} x^m < 1$ for x close to 0, (8.32) holds.   Q.E.D.

Proof of Theorem 8.2.3: Let $\Psi_\pm = \Psi_\pm(x, y)$ be the functions defined in (8.28). By Lemma 8.2.4, $\Psi_\pm$ satisfy (8.32), i.e.,

$$(\Psi_\pm)_{x^k} = \Psi_\pm (\Psi_\pm)_{y^k} \qquad (8.33)$$

with $\Psi_\pm(0, y) = \phi_\pm(y) = \varphi(y) \pm \psi(y)$. It follows from (8.33) that

$$(\Psi_\pm)_{x^k y^l} y^k = (\Psi_\pm)_{x^l}.$$

This implies that $F := \frac{1}{2}\{\Psi_+ - \Psi_-\}$ satisfies (3.18). Hence F is projectively flat by Theorem 3.3.1. Observe that

$$F_{x^k} y^k = \frac{1}{2}\left\{\Psi_+ (\Psi_+)_{y^k} - \Psi_- (\Psi_-)_{y^k}\right\} y^k = \frac{1}{2}\left\{\Psi_+^2 - \Psi_-^2\right\}.$$

Thus the projective factor $P = \frac{1}{2}F^{-1}F_{x^k}y^k$ is given by

$$P := \frac{F_{x^k} y^k}{2F} = \frac{1}{2}\frac{\Psi_+^2 - \Psi_-^2}{\Psi_+ - \Psi_-} = \frac{1}{2}\left\{\Psi_+ + \Psi_-\right\}.$$

By a similar argument, one obtains

$$P^2 - P_{x^k} y^k = -\left(\frac{\Psi_+ - \Psi_-}{2}\right)^2 = -F^2.$$

Thus the flag curvature $\mathbf{K} = -1$ by (6.8).   Q.E.D.

By the formulas in Theorem 8.2.3, one can construct some projectively flat Finsler metrics with $\mathbf{K} = -1$.

Example 8.2.5 Let $\phi = \phi(y)$ be a Minkowski norm on $\mathbf{R}^n$ and $\mathcal{U} :=\{y \in \mathbf{R}^n \mid \phi(y) < 1\}$. Let

$$\psi := \frac{\phi(y) - \delta\phi(\varepsilon y)}{2}, \qquad \varphi := \frac{\phi(y) + \delta\phi(\varepsilon y)}{2},$$

where $\varepsilon = \pm 1$ and δ is a constant chosen so that ψ is a Minkowski norm on $\mathbf{R}^n$. Note that

$$\varphi(y) + \psi(y) = \phi(y), \qquad \varphi(y) - \psi(y) = \delta\phi(\varepsilon y).$$

Let $\Psi_+ = \Psi_+(x, y)$ and $\Psi_- = \Psi_-(x, y)$ be the solutions of (8.28) with $\phi_+(y) = \phi(y)$ and $\phi_-(y) := \delta\phi(\varepsilon y)$, respectively. Let $\Theta = \Theta(x, y)$ denote the Funk metric of ϕ defined in (8.26). Then

$$\Psi_+(x, y) = \Theta(x, y), \qquad \Psi_-(x, y) = \delta\Theta(\delta\varepsilon x, \varepsilon y).$$

By Theorem 8.2.3, one concludes that the following function

$$F := \frac{1}{2}\Big\{\Theta(x, y) - \delta\Theta(\delta\varepsilon x, \varepsilon y)\Big\}$$

is a projectively flat Finsler metric on its domain with $\mathbf{K} = -1$ and its projective factor is given by

$$P = \frac{1}{2}\Big\{\Theta(x, y) + \delta\Theta(\delta\varepsilon x, \varepsilon y)\Big\}.$$

When $\delta = -1$ and $\varepsilon = -1$,

$$\Psi_+(x, y) = \Theta(x, y), \qquad \Psi_-(x, y) = -\Theta(x, -y).$$

By the above argument, we know that

$$F := \frac{1}{2}\Big\{\Theta(x, y) + \Theta(x, -y)\Big\}$$

is a projectively flat Finsler metric on $\mathcal{U}$ with $\mathbf{K} = -1$ and its projective factor is given by

$$P = \frac{1}{2}\Big\{\Theta(x, y) - \Theta(x, -y)\Big\}.$$

F is the Hilbert metric on $\mathcal{U}$ (see Example 3.4.5).

We now consider the special case when $\phi = |y| + \langle a, y \rangle$ where $a \in \mathbf{R}^n$ with $|a| < 1$. The Funk metric defined by ϕ is a Randers metric $\Theta =$

$\alpha(x, y) + \beta(x, y)$, where $\alpha = \alpha(x, y)$ and $\beta = \beta(x, y)$ are given by

$$\alpha := \frac{\sqrt{\left[(1 - \langle a, x \rangle)^2 - |x|^2\right]\left[|y|^2 - \langle a, y \rangle^2\right] + \left[(1 - \langle a, x \rangle)\langle a, y \rangle + \langle x, y \rangle\right]^2}}{(1 - \langle a, x \rangle)^2 - |x|^2},$$

$$\beta := \frac{(1 - \langle a, x \rangle)\langle a, y \rangle + \langle x, y \rangle}{(1 - \langle a, x \rangle)^2 - |x|^2}.$$

Let $\varepsilon = \pm 1$ and δ be a constant with $\delta < 1$ and $|1 - \varepsilon\delta||a| < 1 - \delta$. Then

$$F := \frac{1}{2}\left\{\alpha(x, y) - \delta\alpha(\delta\varepsilon x, y)\right\} + \frac{1}{2}\left\{\beta(x, y) - \delta\varepsilon\beta(\delta\varepsilon x, y)\right\}$$

is a projectively flat Finsler metric near the origin with $\mathbf{K} = -1$ and its projective factor is given by

$$P = \frac{1}{2}\left\{\alpha(x, y) + \delta\alpha(\delta\varepsilon x, y)\right\} + \frac{1}{2}\left\{\beta(x, y) + \delta\varepsilon\beta(\delta\varepsilon x, y)\right\}.$$

Note that F is no longer of Randers type.

Example 8.2.6 Let $\phi = \phi(y)$ be an arbitrary Minkowski norm on $\mathbb{R}^n$ and $\Theta = \Theta(x, y)$ denote the Funk metric of ϕ. For a constant vector $a \in \mathbb{R}^n$, let

$$\psi := \frac{1}{2}\left(\phi(y) + \langle a, y \rangle\right), \qquad \varphi := \frac{1}{2}\left(\phi(y) - \langle a, y \rangle\right),$$

such that

$$\varphi(y) + \psi(y) = \phi(y), \qquad \varphi(y) - \psi(y) = -\langle a, y \rangle.$$

Let $\Psi_+ = \Psi_+(x, y)$ and $\Psi_- = \Psi_-(x, y)$ be defined by (8.28) with $\phi_+(y) = \phi(y)$ and $\phi_-(y) = -\langle a, y \rangle$, respectively. Then

$$\Psi_+ = \Theta(x, y), \qquad \Psi_- = -\frac{\langle a, y \rangle}{1 + \langle a, x \rangle}.$$

By Theorem 8.2.3, one knows that the following function

$$F = \frac{1}{2}\left\{\Theta(x, y) + \frac{\langle a, y \rangle}{1 + \langle a, x \rangle}\right\} \tag{8.34}$$

is projectively flat with $\mathbf{K} = -1$ and its projective factor is given by

$$P = \frac{1}{2}\left\{\Theta(x, y) - \frac{\langle a, y \rangle}{1 + \langle a, x \rangle}\right\}.$$

When $\phi = |y|$, the Funk metric Θ on the unit ball $B^n(1)$ is given by (8.27). Thus for any constant vector $a \in \mathbf{R}^n$ with $|a| < 1$, the following function

$$F(x,y) = \frac{1}{2}\left\{ \frac{\sqrt{|y|^2 - (|x|^2|y|^2 - \langle x,y\rangle^2)} + \langle x,y\rangle}{1 - |x|^2} + \frac{\langle a,y\rangle}{1 + \langle a,x\rangle} \right\}$$

is projectively flat on $B^n(1)$ with $\mathbf{K} = -1$. See Examples 5.1.3 and 6.1.5 above.

Now we are going to construct a projectively flat Finsler metric of zero flag curvature for any given pair $\{\varphi, \psi\}$.

Theorem 8.2.7 ([89]) *Let $\psi = \psi(y)$ be a Minkowski norm on $\mathbf{R}^n$ and $\varphi = \varphi(y)$ be a positively homogeneous function of degree one on $\mathbf{R}^n$. Define $P = P(x,y)$ by*

$$P(x,y) = \varphi\Big(y + P(x,y)x\Big). \tag{8.35}$$

Let

$$F := \psi\Big(y + P(x,y)\, x\Big)\Big\{1 + P_{y^m}(x,y)x^m\Big\}. \tag{8.36}$$

Then $F = F(x,y)$ is a projectively flat Finsler metric of zero flag curvature with $F(0,y) = \psi(y)$ and its projective factor is $P = P(x,y)$ with $P(0,y) = \varphi(y)$.

Proof: By Lemma 8.2.4, P satisfies

$$P_{x^k} = PP_{y^k} \tag{8.37}$$

with $P(0,y) = \varphi(y)$. It follows from (8.37) that

$$P_{x^k y^m} = \frac{1}{2}(P^2)_{y^k y^m} = P_{x^m y^k}. \tag{8.38}$$

Differentiating (8.36) with respect to x^k and using (8.37), one obtains

$$F_{x^k} = \psi_{y^m}(y + Px)\Big\{P\delta_k^m + P_{x^k}x^m\Big\}\Big\{1 + P_{y^l}x^l\Big\}$$
$$+ \psi(y + Px)\Big\{P_{y^k} + P_{y^m x^k}x^m\Big\}.$$

By (8.37), one obtains

$$PF = \psi(y + Px)\Big\{P + P_{x^m}x^m\Big\}. \tag{8.39}$$

Differentiating (8.39) with respect to y^k yields

$$(PF)_{y^k} = \psi_{y^m}(y + Px)\left\{\delta_k^m + P_{y^k}x^m\right\}\left\{P + P_{x^l}x^l\right\}$$
$$+ \psi(y + Px)\left\{P_{y^k} + P_{x^m y^k}x^m\right\}.$$

By (8.37) again, one obtains

$$\left\{P\delta_k^m + P_{x^k}x^m\right\}\left\{1 + P_{y^l}x^l\right\} = \left\{\delta_k^m + P_{y^k}x^m\right\}\left\{P + P_{x^l}x^l\right\}.$$

Together with (8.38), one can see that F satisfies (8.24). Note that (8.37) is equivalent to (8.25) with $\lambda = 0$. Thus F is a projectively flat Finsler metric on its domain with $\mathbf{K} = 0$ and its projective factor is P.   Q.E.D.

By Theorem 8.2.7, one can construct several projectively flat Finsler metrics with $\mathbf{K} = 0$.

Example 8.2.8 Let $\phi = \phi(y)$ be a Minkowski norm on $\mathbf{R}^n$ and $\Theta = \Theta(x, y)$ denote the Funk metric of ϕ defined in (8.26). Let $\psi := \phi(y) + \langle a, y \rangle$ and $\varphi := \phi(y)$. Let $P = P(x, y)$ and $F = F(x, y)$ be defined in (8.35) and (8.36) respectively. Clearly, $P = \Theta$. Observe that

$$\phi(y + Px) = \phi(y + \Theta x) = \Theta.$$

Thus

$$\psi(y + Px) = \Theta + \langle a, y \rangle + \langle a, x \rangle \Theta.$$

The function F defined in (8.36) is given by

$$F = \left\{\left(1 + \langle a, x \rangle\right)\Theta(x, y) + \langle a, y \rangle\right\}\left\{1 + \Theta_{y^k}(x, y)x^k\right\}. \tag{8.40}$$

By Theorem 8.2.7, one knows that F is projectively flat with $\mathbf{K} = 0$ and its projective factor $P = \Theta(x, y)$. Note that when $a = 0$, the Finsler metric in (8.40) is reduced to

$$F = \Theta(x, y)\left\{1 + \Theta_{y^k}(x, y)x^k\right\} = \Theta(x, y) + \Theta_{x^k}(x, y)x^k.$$

When $\psi = |y| + \langle a, y \rangle$ and $\varphi = |y|$, the Finsler metric in (8.40) can be expressed by

$$F = \left\{1 + \langle a, x \rangle + \frac{(1 - |x|^2)\langle a, y \rangle}{\sqrt{|y|^2 - (|x|^2|y|^2 - \langle x, y \rangle^2)} + \langle x, y \rangle}\right\}$$

$$\times \frac{\left(\sqrt{|y|^2 - (|x|^2|y|^2 - \langle x, y \rangle^2)} + \langle x, y \rangle \right)^2}{(1 - |x|^2)^2 \sqrt{|y|^2 - (|x|^2|y|^2 - \langle x, y \rangle^2)}}. \tag{8.41}$$

Clearly, F is not locally Minkowskian. When $a = 0$, the Finsler metric in (8.41) is reduced to

$$F = \frac{\left(\sqrt{|y|^2 - (|x|^2|y|^2 - \langle x, y \rangle^2)} + \langle x, y \rangle \right)^2}{(1 - |x|^2)^2 \sqrt{|y|^2 - (|x|^2|y|^2 - \langle x, y \rangle^2)}}. \tag{8.42}$$

This is just the projectively flat Finsler metric constructed by L. Berwald [17]. The Finsler metric F in (8.42) is positively complete, i.e., every unit speed geodesic on (δ, τ) can be extended to a geodesic on (δ, ∞).

The Finsler metric defined in (8.42) is positively complete, but not complete. To construct a complete one, one should construct a reversible metric $F = F(x, y)$ by (8.35) and (8.36). It suffices to assume that $\varphi = \varphi(y)$ is homogeneous of degree one $(\varphi(\lambda y) = \lambda \varphi(y))$ and $\psi = \psi(y)$ is a Minkowski norm on $\mathbb{R}^n$. A natural question arises: are there φ and ψ with the above properties such that the Finsler metric defined in (8.35) and (8.36) is complete? As a matter of fact, it has been shown that any complete projectively flat Finsler metric with zero flag curvature $\mathbf{K} = 0$ must be a Minkowski metric (see [16] and [39]).

Now we discuss the positive curvature case. Given a Minkowski norm, $\psi = \psi(y)$, and a positively homogeneous function of degree one, $\varphi = \varphi(y)$, on $\mathbb{R}^n$, one would like to find a projectively flat Finsler metric $F = F(x, y)$ of constant flag curvature $\mathbf{K} = 1$ satisfying that $F(0, y) = \psi(y)$ and its projective factor $P(0, y) = \varphi(y)$. This problem turns out to be very difficult.

Suppose that a complex-valued function $H = H(x, y)$ satisfies the following system

$$H_{x^k} = H H_{y^k} \tag{8.43}$$

with $H(0, y) = \varphi(y) + i\psi(y)$. Express $H = P + iF$ with $P(0, y) = \varphi(y)$ and $F(0, y) = \psi(y)$. By continuity, $F = F(x, y)$ is a Finsler metric for x close to the origin. It follows from (8.43) that

$$P_{x^k} - P P_{y^k} + F F_{y^k} + i \left\{ F_{x^k} - (PF)_{y^k} \right\} = 0.$$

That is,

$$F_{x^k} = (PF)_{y^k}, \qquad P_{x^k} = PP_{y^k} - FF_{y^k}.$$

By Lemma 8.2.1, $F = F(x, y)$ is projectively flat with constant flag curvature $\mathbf{K} = 1$ and its projective factor is $P = P(x, y)$. Thus, the remaining problem is how to find complex-valued functions $H = H(x, y)$ satisfying (8.43) with $H(0, y) = \varphi(y) + i\psi(y)$. We might construct such functions using a Taylor expansion.

Let $H = H(x, y)$ be a positively homogeneous function of degree one satisfying (8.43) with $H(0, y) = \varphi(y) + i\psi(y)$. By (8.43), we have

$$H_{x^{i_1} \cdots x^{i_m}} = \frac{1}{m+1} \Big[H^{m+1} \Big]_{y^{i_1} \cdots y^{i_m}}.$$

Thus

$$H_{x^{i_1} \cdots x^{i_m}}(0, y) = \frac{1}{m+1} \Big[(\varphi + i\psi)^{m+1} \Big]_{y^{i_1} \cdots y^{i_m}}(y).$$

If $H = H(x, y)$ is analytic in x at $x = 0$, then

$$H = \sum_{m=0}^{\infty} \frac{1}{(m+1)!} \frac{d^m}{dt^m} \Big[\big(\varphi(y + tx) + i\psi(y + tx) \big)^{m+1} \Big]|_{t=0}. \qquad (8.44)$$

However, for a given pair $\{\varphi, \psi\}$, it is not easy to check the convergence of the power series in (8.44).

Here is another possible way to construct a particular solution of (8.43) with $H(0, y) = \varphi(y) + i\psi(y)$. Assume that $\phi := \varphi(y) + i\psi(y)$ can be extended to a function $\tilde{\phi} := \phi(y + zx)$, $z \in \mathbb{C}$ such that for any x close to the origin and any $y \in \mathbb{R}^n$, there is a complex number z satisfying

$$z = \phi(y + zx). \qquad (8.45)$$

Then the solution $H := z$ satisfies (8.43) with $H(0, y) = \varphi(y) + i\psi(y)$. This method can be employed to construct a family of projectively flat Finsler metrics on $\mathbb{R}^n$ with $\mathbf{K} = 1$.

Example 8.2.9 Let $\psi := \cos(\alpha)|y|$ and $\varphi := \sin(\alpha)|y|$ where $|\cdot|$ denotes the Euclidean norm on $\mathbb{R}^n$ and α is an angle with $|\alpha| < \pi/2$. Then

$$\phi = \varphi(y) + i\psi(y) = ie^{-i\alpha}|y|.$$

Note that $|y + zx| = \sqrt{\langle y + zx, y + zx \rangle}$ can be defined for $z \in \mathbb{C}$. Equation (8.45) becomes

$$z = ie^{-i\alpha} \sqrt{|y|^2 + 2\langle x, y \rangle z + |x|^2 z^2}.$$

One obtains a formula for $H := z$,

$$H = \frac{-\langle x, y \rangle + i\sqrt{(e^{2i\alpha} + |x|^2)|y|^2 - \langle x, y \rangle^2}}{e^{2i\alpha} + |x|^2}. \tag{8.46}$$

From (8.46), one can see that F is a complex-valued function on $\times \mathbb{R}^n$. We can express it in the form $H := P + iF$. By the above argument, $F = F(x, y)$ is a projectively flat Finsler metric of constant flag curvature $\mathbf{K} = 1$ and $P = P(x, y)$ is the projective factor of F. To express F and P in terms of elementary functions, let

$$A : = \left(\cos(2\alpha)|y|^2 + (|x|^2|y|^2 - \langle x, y \rangle^2) \right)^2 + \left(\sin(2\alpha)|y|^2 \right)^2,$$
$$B : = \cos(2\alpha)|y|^2 + (|x|^2|y|^2 - \langle x, y \rangle^2),$$
$$U : = \sin(2\alpha)\langle x, y \rangle,$$
$$V : = \left(\cos(2\alpha) + |x|^2 \right)\langle x, y \rangle,$$
$$E : = |x|^4 + 2\cos(2\alpha)|x|^2 + 1.$$

For an angle α with $0 \leq \alpha < \pi/2$,

$$\sqrt{(e^{2\alpha i} + |x|^2)|y|^2 - \langle x, y \rangle^2} = \sqrt{\frac{\sqrt{A} + B}{2}} + i\sqrt{\frac{\sqrt{A} - B}{2}}.$$

From (8.46), one obtains

$$F = \sqrt{\frac{\sqrt{A} + B}{2E} + \left(\frac{U}{E} \right)^2} + \frac{U}{E}, \tag{8.47}$$

$$P = \sqrt{\frac{\sqrt{A} - B}{2E} - \left(\frac{U}{E} \right)^2} - \frac{V}{E}. \tag{8.48}$$

In dimension two, one can verify that the Finsler metric in (8.47) is a Bryant metric [19], [20].

According to (8.44), one can also express H as a power series,

$$H = \sum_{m=0}^{\infty} \frac{1}{(m+1)!} \frac{d^m}{dt^m} \left[\left(ie^{-i\alpha} |y + tx| \right)^{m+1} \right]_{|t=0}.$$

Thus F and P can be expressed as power series,

$$F = \sum_{m=0}^{\infty} \frac{\sin[(m+1)(\pi/2 - \alpha)]}{(m+1)!} \frac{d^m}{dt^m} \left[|y + tx|^{m+1} \right]_{|t=0}$$

$$P = \sum_{m=0}^{\infty} \frac{\cos[(m+1)(\pi/2 - \alpha)]}{(m+1)!} \frac{d^m}{dt^m} \left[|y + tx|^{m+1} \right]_{|t=0}.$$

Projectively flat Finsler metrics of constant flag curvature $\mathbf{K} = 1$ constructed above are all local in a sense that they are defined on an open convex domain in $\mathbf{R}^n$ and hence they are incomplete. The Finsler metrics in Example 8.2.9 can be pulled back to S^n by two maps $\psi_\pm$ to form complete irreversible projectively flat Finsler metrics of constant flag curvature $\mathbf{K} = 1$. The maps $\psi_\pm$ are defined in Example 1.2.4.

Recently, R. Bryant has completely determined the global structures of Finsler metrics with $\mathbf{K} = 1$ on S^n, whose geodesics are great circles. His approach is very nice and different from the above one. See Bryant's work [19] and [20], [21].

8.3   Projectively Flat Metrics with Almost Isotropic S-Curvature

In this section we are going to study and characterize locally projectively flat Finsler metrics with almost isotropic S-curvature. Recall that a Finsler metric $F = F(x, y)$ is said to have almost isotropic S-curvature, if

$$\mathbf{S} = (n+1)\left\{ cF + \eta \right\}, \tag{8.49}$$

where $c = c(x)$ is a scalar function and $\eta = \eta_i(x)y^i$ is a closed 1-form on M. We know that a Randers metric has isotropic S-curvature if and only if it has almost isotropic S-curvature. In Proposition 8.1.2, we have completely classified projectively flat Randers metrics with isotropic S-curvature.

According to Example 5.1.3, the following Finsler metric F is projectively flat with constant flag curvature and almost isotropic S-curvature,

$$F := \Theta(x, y) + \frac{\langle a, y \rangle}{1 + \langle a, x \rangle}, \qquad y \in T_x \mathcal{U} \cong \mathrm{R}^n,$$

where $\Theta = \Theta(x, y)$ is the Funk metric on a strongly convex domain $\mathcal{U} \subset \mathrm{R}^n$. When $\mathcal{U}$ is the standard unit ball $\mathrm{B}^n(1)$, F is of isotropic S-curvature. A natural problem arises: are there other types of projectively flat Finsler metrics of almost isotropic S-curvature. The answer is that if the metric is not a Randers metric, then, after a scaling, it must be in the above form or its reversed form. More precisely, we have the following

Proposition 8.3.1 ([26]) *Let $F = F(x, y)$ be a projectively flat Finsler metric on an open subset $\mathcal{U} \subset \mathrm{R}^n$. Suppose that F has almost isotropic S-curvature, i.e.,*

$$\mathbf{S} = (n + 1)\Big\{cF + \eta\Big\}, \tag{8.50}$$

where $c = c(x)$ is a scalar function and $\eta = \eta_i dx^i$ is a closed 1-form on $\mathcal{U}$. Then F is determined as follows.

(a) *If $\mathbf{K} \neq -c(x)^2 + \frac{c_{x^m}(x)y^m}{F(x,y)}$ at every point $x \in \mathcal{U}$, then $F = \alpha + \beta$ is a Randers metric on $\mathcal{U}$ and it is determined in Proposition 8.1.2 (B);*
(b) *If $\mathbf{K} \equiv -c(x)^2 + \frac{c_{x^m}(x)y^m}{F(x,y)}$, then $c(x) = c$ is a constant, and either F is locally Minkowskian ($c = 0$) or up to a scaling, F can be expressed as*

$$F = \begin{cases} \Theta(x, y) + \frac{\langle a, y \rangle}{1 + \langle a, x \rangle} & (c = \frac{1}{2}) \\ \Theta(x, -y) - \frac{\langle a, y \rangle}{1 + \langle a, x \rangle} & (c = -\frac{1}{2}) \end{cases}, \tag{8.51}$$

where $a \in \mathrm{R}^n$ is a constant vector and $\Theta = \Theta(x, y)$ is a Funk metric defined by (1.38).

Proof. By assumption, F is projectively flat. Thus the spray coefficients are given by $G^i = Py^i$, where

$$P := \frac{F_{x^k}y^k}{2F}. \tag{8.52}$$

By (6.8), the flag curvature of F is given by

$$\mathbf{K} = \frac{P^2 - P_{x^k} y^k}{F^2}. \tag{8.53}$$

The S-curvature is given by

$$\mathbf{S} = (n+1)\Big\{ cF + \eta \Big\}$$

$$= \frac{\partial G^m}{\partial y^m} - y^m \frac{\partial (\ln \sigma_F)}{\partial y^m}$$

$$= (n+1)P - y^m \frac{\partial (\ln \sigma_F)}{\partial y^m}.$$

Since $\eta = \eta(x,y)$ is closed on $\mathcal{U}$, it can be expressed in the form $\eta = dh_x(y)$ where $h = h(x)$ is a scalar function on $\mathcal{U}$. From the above identities, we obtain

$$P = cF + d\varphi, \tag{8.54}$$

where $\varphi := h(x) + \frac{1}{n+1} \ln[\sigma_F(x)]$. It follows from (8.52) and (8.54) that

$$F_{x^m} y^m = 2FP = 2F\Big\{ cF + \varphi_{x^m} y^m \Big\}. \tag{8.55}$$

Substituting (8.54) into (8.53) and using (8.55), one obtains

$$\mathbf{K} = \frac{\Big\{ cF + \varphi_{x^m} y^m \Big\}^2 - \Big\{ c_{x^m}(x) y^m F + cF_{x^m} y^m + \varphi_{x^i x^j} y^i y^j \Big\}}{F^2}$$

$$= \frac{-c^2 F^2 - c_{x^m} y^m F + \Big\{ \varphi_{x^i} \varphi_{x^j} - \varphi_{x^i x^j} \Big\} y^i y^j}{F^2}. \tag{8.56}$$

On the other hand, since F is of scalar flag curvature, by Proposition 7.1.2, the flag curvature of F is given by

$$\mathbf{K} = 3\frac{c_{x^m} y^m}{F} + \sigma, \tag{8.57}$$

where $\sigma = \sigma(x)$ is a scalar function on $\mathcal{U}$. Comparing (8.56) with (8.57) yields

$$\Big\{ \sigma + c^2 \Big\} F^2 + 4c_{x^m} y^m F + \Big\{ \varphi_{x^i x^j} - \varphi_{x^i} \varphi_{x^j} \Big\} y^i y^j = 0. \tag{8.58}$$

Assume that $\mathbf{K} \neq -c(x)^2 + \frac{c_{x^m}(x)y^m}{F(x,y)}$ at every point $x \in \mathcal{U}$. Then, by (8.57), for any $x \in \mathcal{U}$, there is a non-zero vector $y \in T_x\mathcal{U}$ such that

$$\sigma(x) + c(x)^2 + \frac{2c_{x^m}(x)y^m}{F(x,y)} \neq 0.$$

We claim that $\sigma(x) + c(x)^2 \neq 0$ for any $x \in \mathcal{U}$. If not, there is a point $x_o \in \mathcal{U}$ such that $\sigma(x_o) + c(x_o)^2 = 0$. The above inequality implies that $dc \neq 0$ at x_o. Then (8.58) at x_o is reduced to

$$4c_{x^m}(x_o)y^m F(x_o, y) + \left\{ \varphi_{x^i x^j}(x_o) - \varphi_{x^i}(x_o)\varphi_{x^j}(x_o) \right\} y^i y^j = 0.$$

$\phi := F(x_o, y)$ is the so-called *Kropina* metric, hence singular in certain direction $y \in T_{x_o}M$. Such metrics are not under our consideration.

Now we may assume that $\sigma(x) + c(x)^2 \neq 0$ for any $x \in \mathcal{U}$. One can solve the quadratic equation (8.58) for F,

$$F = \frac{\sqrt{[\sigma + c^2][\varphi_{x^i x^j} - \varphi_{x^i}\varphi_{x^j}]y^i y^j + 4[c_{x^m}y^m]^2} - 2c_{x^m}y^m}{\sigma + c^2}.$$

That is, $F = \alpha + \beta$ is a Randers metric. In this case, by Lemma 5.2.2, $\mathbf{S}$ is isotropic, i.e., $\eta = 0$. Since F is projectively flat, α is of constant sectional curvature $\mathbf{K}_\alpha = \mu$ and β is closed. Moreover, by Proposition 8.1.2, the flag curvature is given by (8.57) with $\sigma = 3c(x)^2 + \mu$. See (8.11). Note that the inequality $\sigma(x) + c(x)^2 \neq 0$ is equivalent to the following inequality

$$\mu + 4c(x)^2 \neq 0.$$

By Proposition 8.1.2 (B), F is given by (8.13).

We now assume that $\mathbf{K} \equiv -c(x)^2 + \frac{c_{x^i}(x)y^i}{F(x,y)}$. It follows from (8.57) that

$$\sigma(x) + c(x)^2 + \frac{2c_{x^m}(x)y^m}{F} \equiv 0.$$

This implies that $c(x) = c$ is a constant, hence $\sigma(x) = -c^2$ is a constant too. In this case, the flag curvature is given by $\mathbf{K} = -c^2$. The equation (8.58) is reduced to

$$\varphi_{x^i x^j} - \varphi_{x^i}\varphi_{x^j} = 0.$$

It is easy to solve this equation,

$$\varphi = -\ln\left(1 + \langle a, x\rangle\right) + C,$$

where $a \in \mathbf{R}^n$ is a constant vector and C is a constant.

If $c = 0$, then $\mathbf{K} = -c^2 = 0$. It follows from (8.54) that the projective factor $P = d\varphi$ is a 1-form, hence the spray coefficients $G^i = Py^i$ are quadratic in $y \in T_x\mathcal{U}$. By definition, F is a Berwald metric. One concludes that F is locally Minkowskian by Theorem 2.3.2.

If $c \neq 0$, then $\mathbf{K} = -c^2 < 0$.

$$\Psi := P + cF = 2cF + d\varphi.$$

Then

$$F = \frac{1}{2c}\Big\{\Psi(x,y) - d\varphi_x(y)\Big\} = \frac{1}{2c}\Big\{\Psi(x,y) + \frac{\langle a,y\rangle}{1 + \langle a,x\rangle}\Big\}.$$

It follows from Lemma 8.2.1 that

$$F_{x^k} = (PF)_{y^k}, \qquad P_{x^k} = PP_{y^k} + c^2FF_{y^k}.$$

We have

$$\begin{aligned}
\Psi_{x^i} &= P_{x^i} + cF_{x^i}\\
&= PP_{y^i} + c^2FF_{y^i} + c(PF)_{y^i}\\
&= \{P + cF\}\{P_{y^i} + cF_{y^i}\} = \Psi\Psi_{y^i}.
\end{aligned}$$

Since c might be negative, F and Ψ have opposite signs when a is sufficiently small. Thus we introduce another function as follows,

$$\Theta := \begin{cases} \Psi(x,y) & \text{if } c > 0\\ -\Psi(x,-y) & \text{if } c < 0 \end{cases}.$$

Then $\Theta = \Theta(x,y)$ satisfies (1.38), i.e.,

$$\Theta_{x^i} = \Theta\Theta_{y^i}.$$

By definition, Θ is a Funk metric. Choosing $c = \pm\frac{1}{2}$, we obtain (8.51).

$$\text{Q.E.D.}$$

Appendix A

Maple Programs

In Finsler geometry, the computations of geometric quantities are usually very complicated. However, using a Maple program, we can quickly check whether or not an expression is correct even though we can derive it manually; we can simplify the expressions in the computation without making mistakes.

A.1 Spray Coefficients of Two-dimensional Finsler Metrics

In this section, we shall compute the spray coefficients for two-dimensional Finsler metrics. For simplicity, we denote a point (x^1, x^2) in R^2 by (x, y) and a tangent vector $y^1 \frac{\partial}{\partial x^1} + y^1 \frac{\partial}{\partial x^2}$ at (x^1, x^2) by $(x, y; u, v)$. Let $F = F(x, y, u, v)$ be a Finsler metric on an open subset $\mathcal{U} \subset \mathrm{R}^2$. Let

$$L(x, y; u, v) := \frac{1}{2} F^2(x, y, u, v).$$

The spray coefficients, $G := G^1(x, y; u, v)$ and $H := G^2(x, y; u, v)$, are given by

$$G := \frac{(L_x L_{vv} - L_y L_{uv}) - (L_{xv} - L_{yu}) L_v}{2 \left(L_{uu} L_{vv} - (L_{uv})^2 \right)} \qquad (A.1)$$

$$H := \frac{(-L_x L_{uv} + L_y L_{uu}) + (L_{xv} - L_{yu}) L_u}{2 \left(L_{uu} L_{vv} - (L_{uv})^2 \right)}. \qquad (A.2)$$

Below is a Maple program for computing G and H. Our testing example is the Klein metric on the unit disk in R^2. It is defined by

$$F = \frac{\sqrt{(u^2 + v^2) - ((x^2 + y^2)(u^2 + v^2) - (xu + yv)^2)}}{1 - (x^2 + y^2)}.$$

It takes a while if one computes the spray coefficients G^i by hands. Now we can use a Maple program to find its spray coefficients within a couple of seconds once the metric function is entered.

```
> restart;
> GH:=proc(F)
> local L,Lx,Ly,Lu,Lv,Lxv,Lyu,Luu,Luv,Lvv,Num1,Num2,
> Den,G,H;
> L:=F^2/2;
> Lx:=diff(L,x);
> Ly:=diff(L,y);
> Lu:=diff(L,u);
> Lv:=diff(L,v);
> Lxv:=diff(L,x,v);
> Lyu:=diff(L,y,u);
> Luu:=diff(L,u,u);
> Luv:=diff(L,u,v);
> Lvv:=diff(L,v,v);
> Den:=2*(Luu*Lvv-Luv*Luv);
> Num1:=(Lx*Lvv-Ly*Luv)-(Lxv-Lyu)*Lv;
> Num2:=(-Lx*Luv+Ly*Luu)+(Lxv-Lyu)*Lu;
> G:=Num1/Den;
> H:=Num2/Den;
> RETURN([G,H]);
> end:
> F:=sqrt(yy-(xx*yy-xy^2))/(1-xx);
```

$$F := \frac{\sqrt{yy - xx\,yy + xy^2}}{1 - xx}$$

```
> xx:=x^2+y^2:
> yy:=u^2+v^2:
```

```
>  xy:=x*u+y*v:
>  M:=GH(F):
>  G:=M[1]:
>  H:=M[2]:
>  G:=simplify(G);
```

$$G := -\frac{u\,(x\,u + y\,v)}{-1 + x^2 + y^2}$$

```
>  H:=simplify(H);
```

$$H := -\frac{v\,(x\,u + y\,v)}{-1 + x^2 + y^2}$$

In the above program, we define a function GH(F). The input is a Finsler metric F and the output is the matrix $[G, H]$ formed by the spray coefficients. For the Klein metric, one gets G and H which can be expressed in the form $G = Pu$ and $H = Pv$. Thus the Klein metric is projectively flat on the unit disk.

The above Maple program can be used to compute the spray coefficients for general two-dimensional Finsler metrics. The Bryant metrics on the upper (or lower) hemisphere can be pulled to projective metrics defined on $\mathbf{R}^2$ using the same standard map $\psi_\pm$ in Example 1.2.4. Thus they can be expressed as metrics on $\mathbf{R}^2$.

$$F := \sqrt{\frac{\sqrt{A} + B}{2E} + \left(\frac{U}{E}\right)^2} + \frac{U}{E},$$

where

$$A := B^2 + (1 - e^2)\,(u^2 + v^2)^2,$$
$$B := e\,(u^2 + v^2) + [(x^2 + y^2)(u^2 + v^2) - (xu + yv)^2],$$
$$U := \sqrt{1 - e^2}\,(xu + yv),$$
$$V := [e + (x^2 + y^2)]\,(xu + yv),$$
$$E := 1 + 2e\,(x^2 + y^2) + (x^2 + y^2)^2.$$

Below is a portion of a Maple program for computing G and H of a family of Bryant metrics on S^2. We omit the head part of the above Maple program defining GH(F).

```
>  F:=sqrt((sqrt(A)+B)/(2*E)+(U/E)^2)+U/E;
```

$$F := \frac{1}{2}\sqrt{2\frac{\sqrt{A}+B}{E}+\frac{4U^2}{E^2}}+\frac{U}{E}$$

```
>  P:=sqrt((sqrt(A)-B)/(2*E)-(U/E)^2)-V/E;
```

$$P := \frac{1}{2}\sqrt{2\frac{\sqrt{A}-B}{E}-\frac{4U^2}{E^2}}-\frac{V}{E}$$

```
>  A:=(e*yy+(xx*yy-xy^2))^2+(1-e^2)*yy^2:
```

```
>  B:=e*yy+(xx*yy-xy^2):
```

```
>  U:=sqrt(1-e^2)*xy:
```

```
>  V:=(e+xx)*xy:
```

```
>  E:=1+2*e*xx+xx^2:
```

```
>  xx:=x^2+y^2:
```

```
>  yy:=u^2+v^2:
```

```
>  xy:=x*u+y*v:
```

```
>  M:=GH(F):
```

```
>  G:=M[1]:
```

```
>  H:=M[2]:
```

```
>  x:=-1/2;y:=-3/5;u:=-1;v:=4/5;e:=1/4;
```

$$x := \frac{-1}{2}$$

$$y := \frac{-3}{5}$$

$$u := -1$$

$$v := \frac{4}{5}$$

$$e := \frac{1}{4}$$

```
>  simplify(G/u-H/v);
```

$$0$$

```
>  simplify(G/u-P);
```

$$0$$

```
>  simplify(H/v-P);
```

$$0$$

In the above Maple prgram, we verified two facts: 1) F is projectively flat, and 2) $G = Pu$ and $H = Pv$ where P is given by

$$P = \sqrt{\frac{\sqrt{A} - B}{2E} - \left(\frac{U}{E}\right)^2} + \frac{V}{E}.$$

One can also use equation (3.18) to check whether a Finsler metric is projectively flat. Equation (3.18) in dimension two is given as follows,

$$F_{xu}u + F_{yu}v = F_x, \qquad F_{xv}u + F_{yv}v = F_y. \tag{A.3}$$

If a Finsler metric $F = F(x, y; u, v)$ satisfies (A.3), then by Theorem 3.3.1, it is projectively flat. In this case, the spray coefficients are in the form $G = Pu$ and $H = Pv$, where

$$P = \frac{F_x u + F_y v}{2F}. \tag{A.4}$$

Below is a Maple program by which we verify that a Bryant metric F in (8.47) satisfies (A.3), hence it is projectively flat. We also compare the function P in (8.48) with the function P in (A.4) using fractional numbers. They are always sufficiently close. Thus we conclude that they are equal to each other. A tip is to select small values for the variables, otherwise the computation will go beyond the capacity of Maple.

```
>   restart;
>   F:=sqrt((sqrt(A)+B)/(2*E)+(U/E)^2)+U/E;
```

$$F := \frac{1}{2}\sqrt{2\,\frac{\sqrt{A}+B}{E} + \frac{4\,U^2}{E^2}} + \frac{U}{E}$$

```
>   P:=sqrt((sqrt(A)-B)/(2*E)-(U/E)^2)-V/E;
```

$$P := \frac{1}{2}\sqrt{2\,\frac{\sqrt{A}-B}{E} - \frac{4\,U^2}{E^2}} - \frac{V}{E}$$

```
>   A:=(e*yy+(xx*yy-xy^2))^2+(1-e^2)*yy^2:
>   B:=e*yy+(xx*yy-xy^2):
>   U:=sqrt(1-e^2)*xy:
>   V:=(e+xx)*xy:
>   E:=1+2*e*xx+xx^2:
>   xx:=x^2+y^2:
>   yy:=u^2+v^2:
```

```
>   xy:=x*u+y*v:
>   simplify(diff(F,x,u)*u+diff(F,y,u)*v-diff(F,x));
```
$$0$$
```
>   simplify(diff(F,x,v)*u+diff(F,y,v)*v-diff(F,y));
```
$$0$$
```
>   PP:=(diff(F,x)*u+diff(F,y)*v)/(2*F):
>   e:=1/4;x:=1;y:=-1/2;u:=1/3;v:=1/5;
```
$$e := \frac{1}{4}$$
$$x := 1$$
$$y := \frac{-1}{2}$$
$$u := \frac{1}{3}$$
$$v := \frac{1}{5}$$
```
>   simplify(PP-P);
```
$$0$$

A.2   Gauss Curvature

The flag curvature in dimension two is called the Gauss curvature. For a Finsler metric $F = F(x, y; u, v)$, the Gauss curvature $\mathbf{K} = \mathbf{K}(x, y; u, v)$ is given by

$$\mathbf{K} := \frac{1}{F^2}\Big\{ 2G_x + 2H_y - G_u^2 - H_v^2 - 2H_u G_v$$
$$-Q_x u - Q_y v + 2GQ_u + 2HQ_v \Big\}, \tag{A.5}$$

where $G = G(x, y; u, v)$ and $H = H(x, y; u, v)$ denote its spray coefficients and $Q = G_u + H_v$. The formula (A.5) is in fact a formula for the Ricci scalar $\mathbf{Ric}$ divided by F^2. In dimension two, the quotient $\mathbf{Ric}/F^2$ is the Gauss curvature.

Below is a Maple program for the Gauss curvature, by which we compute the Gauss curvature of the Funk metric on the unit disk. As we know, the Gauss curvature of the Funk metric is equal to -0.25. However, our PC does not run fast enough to complete the symbolic computation. Thus we randomly select some point (x, y) in the unit disk and a direction (u, v). If we always obtain a value sufficiently close to -0.25, then we can conclude that the Gauss curvature is equal to -0.25 and look for a rigorous proof.

```
>   restart;
>   GC:=proc(F)
>   local L,Lx,Ly,Lu,Lv,Lxv,Lyu,Luu,Luv,Lvv,Num1,Num2,
>   Den,G,H,Gx,Hy,Gu,Gv,Hu,Hv,Q,Qx,Qy,Qu,Qv,M,N,K;
>   L:=F^2/2;
>   Lx:=diff(L,x);
>   Ly:=diff(L,y);
>   Lu:=diff(L,u);
>   Lv:=diff(L,v);
>   Lxv:=diff(L,x,v);
>   Lyu:=diff(L,y,u);
>   Luu:=diff(L,u,u);
>   Luv:=diff(L,u,v);
>   Lvv:=diff(L,v,v);
>   Den:=2*(Luu*Lvv-Luv*Luv);
>   Num1:=(Lx*Lvv-Ly*Luv)-(Lxv-Lyu)*Lv;
>   Num2:=(-Lx*Luv+Ly*Luu)+(Lxv-Lyu)*Lu;
>   G:=Num1/Den;
>   H:=Num2/Den;
>   Gx:=diff(G,x);
>   Hy:=diff(H,y);
>   Gu:=diff(G,u):
>   Gv:=diff(G,v):
>   Hu:=diff(H,u):
>   Hv:=diff(H,v):
>   Q:=Gu+Hv:
>   Qx:=diff(Q,x):
>   Qy:=diff(Q,y):
>   Qu:=diff(Q,u):
```

```
>  Qv:=diff(Q,v):
>  M:=Qx*u+Qy*v:
>  N:=G*Qu+H*Qv:
>  K:=(2*Gx+2*Hy-Gu^2-Hv^2-2*Hu*Gv-M+2*N)/F^2:
>  RETURN(K);
>  end:
>  F:=sqrt(Y-(X*Y-Z^2))/(1-X)+Z/(1-X);
```

$$F := \frac{\sqrt{Y - XY + Z^2}}{1 - X} + \frac{Z}{1 - X}$$

```
>  X:=x^2+y^2:Y:=u^2+v^2:Z:=x*u+y*v:
>  K:=GC(F):
>  x:=0.3:y:=-0.6:u:=2.1:v:=0.5:
>  K:=simplify(K);
```

$$K := -.2499999933$$

In the above program, we define a function K:=GC(F). The input is a Finsler metric F and the output is the Gauss curvature **K**. Note that the first half of the program is for computing the spray coefficients G and H.

For a projective flat metric $F = F(x, y; u, v)$, one can first find the projective factor using (A.4), then use the projective factor to compute the flag curvature by the formula in (6.8), i.e.,

$$\mathbf{K} = \frac{P^2 - P_x u - P_y v}{F^2}.$$

A.3   Spray Coefficients of (α, β)-Metrics

In this section, we shall find a formula for the spray coefficients G^i of an (α, β)-metric in *any* dimension. This formula is given in (3.5) without detailed computation. We shall start with computing g_{ij}, then find a formula for g^{ij}, by which we derive a formula for G^i. In each step, we introduce some new variables which are expressed in terms of previous variables. We leave them without simplification until we obtain a formula for G^i. Then we simplify all the coefficients involved in the formula of G^i using Maple.

Let

$$F = \alpha\phi(s), \qquad s = \frac{\beta}{\alpha}.$$

We have

$$g_{ij} = \rho a_{ij} + \rho_0 b_i b_j + \rho_1(b_i \alpha_j + b_j \alpha_i) + \rho_2 \alpha_i \alpha_j, \qquad (A.6)$$

where $\alpha_i := a_{ij} y^j / \alpha$ and

$$\rho := \phi\left\{\phi - s\phi'\right\},$$
$$\rho_0 := \phi\phi'' + \phi'\phi',$$
$$\rho_1 := -\left\{s(\phi\phi'' + \phi'\phi') - \phi\phi'\right\},$$
$$\rho_2 := s\left\{s(\phi\phi'' + \phi'\phi') - \phi\phi'\right\}.$$

Rewrite g_{ij} as follows:

$$g_{ij} = \rho\left\{A_{ij} + \mu\, Y_i Y_j\right\} \qquad (A.7)$$

where

$$Y_i := \alpha_i + \varepsilon b_i,$$
$$A_{ij} := a_{ij} + \delta\, b_i b_j,$$
$$\varepsilon := \frac{\rho_1}{\rho_2},$$
$$\delta := \frac{\rho_0 - \epsilon^2 \rho_2}{\rho},$$
$$\mu := \frac{\rho_2}{\rho}.$$

The inverse $(A^{ij}) := (A_{ij})^{-1}$ is given by

$$A^{ij} = a^{ij} - \tau\, b^i b^j,$$

where

$$\tau = \frac{\delta}{1 + \delta b^2}.$$

Using the formula for A^{ij}, we can find a formula for the inverse $(g^{ij}) := (g_{ij})^{-1}$.

$$g^{ij} = \rho^{-1}\left\{A^{ij} - \eta Y^i Y^j\right\} = \rho^{-1}\left\{a^{ij} - \tau\, b^i b^j - \eta\, Y^i Y^j\right\},$$

where

$$Y^i := A^{ij} Y_j = \frac{y^i}{\alpha} + \lambda\, b^i,$$

$$\lambda := -\tau s + \varepsilon - \varepsilon\tau b^2 = \frac{\varepsilon - \delta s}{1 + \delta b^2}$$

$$\eta := \frac{\mu}{1 + Y^2 \mu},$$

$$Y := \sqrt{Y^i A_{ij} Y^j} = \sqrt{1 + (\lambda + \epsilon)s + \lambda\epsilon b^2}.$$

The geodesic coefficients G^i are given by

$$G^i = G^i_\alpha + \frac{F_{;k} y^k}{2F} y^i + \frac{F}{2} g^{il}\left\{F_{;k\cdot l} y^k - F_{;l}\right\},$$

where $F_{;i}$ denote the covariant derivatives of F with respect to α and $F_{;i\cdot j} = [F_{;i}]_{y^j}$. We use the above identity to find a formula for G^i.

Observe that

$$F_{;k} = \phi' b_{i;k} y^i$$

$$F_{;k\cdot l} y^k = (b_l - s\alpha_l)\phi'' \frac{r_{00}}{\alpha} + \phi' b_{l;k} y^k.$$

Thus

$$F_{;k} y^k = \phi' b_{i;k} y^i y^k = \phi' r_{00}.$$

$$F_{;k\cdot l} y^k - F_{;l} = (b_l - s\alpha_l)\phi'' \frac{r_{00}}{\alpha} + 2\phi' s_{l0}.$$

Then we obtain the following formula for G^i.

$$G^i = G^i_\alpha + P y^i + Q^i,$$

where

$$P = \Xi s_0 + \alpha^{-1}\Theta r_{00}$$

$$Q^i = Q\alpha s^i{}_0 + \left(\Phi\alpha s_0 + \Psi r_{00}\right) b^i$$

where

$$Q = \frac{\phi\phi'}{\rho},$$

$$\Xi = -\frac{\eta\lambda}{\rho}\phi\phi'$$

$$\Theta = \frac{\phi'}{2\phi} - \frac{s\phi\phi''}{2\rho} - \frac{(b^2 - s^2)\eta\lambda\phi\phi''}{2\rho},$$

$$\Phi = -\frac{\phi\phi'}{\rho}(\tau + \eta\lambda^2)$$

$$\Psi = \frac{\phi\phi''}{2\rho}\left\{1 - (b^2 - s^2)(\tau + \eta\lambda^2)\right\}.$$

We use Maple to simplify Q and Θ as follows,

$$Q = \frac{\phi'}{\phi - s\phi'}$$

$$\Theta = \frac{\phi\phi' - s(\phi\phi'' + \phi'\phi')}{2\phi\left((\phi - s\phi') + (b^2 - s^2)\phi''\right)}.$$

We also use Maple to find the following identity,

$$\frac{\Xi}{\Theta} = \frac{\Phi}{\Psi} = -2Q.$$

Let

$$\chi := \frac{\Psi}{\Theta}.$$

χ is given by

$$\chi = \frac{\phi\phi''}{\phi\phi' - s(\phi\phi'' + \phi'\phi')}.$$

Finally, we obtain the following formula for G^i.

$$G^i = G^i_\alpha + Qs^i{}_0 + \Theta\left\{-2Qs_0 + r_{00}\right\}\left\{\frac{y^i}{\alpha} + \chi b^i\right\}.$$

Below is a Maple program for the above computation.

```
>   restart;
>   f:=phi(s):
>   fs:=diff(f,s):
```

```
>   fss:=diff(f,s,s):


>   rho:=f*(f-s*fs):
>   rho0:=f*fss+fs^2:
>   rho1:=-(s*(f*fss+fs^2)-f*fs):
>   rho2:=s*(s*(f*fss+fs^2)-f*fs):


>   epsilon:=rho1/rho2:
>   delta:=(rho0-epsilon^2*rho2)/rho:
>   mu:=rho2/rho:
>   tau:=delta/(1+delta*b2):
>   lambda:=-tau*s+epsilon-epsilon*tau*b2:
>   Y2:=1+(lambda+epsilon)*s+lambda*epsilon*b2:
>   eta:=mu/(1+Y2*mu):


>   Xi:=-(eta*lambda/rho)*f*fs:
>   Theta1:=fs/(2*f)-s*f*fss/(2*rho):
>   Theta2:=-(b2-s^2)*eta*lambda*f*fss/(2*rho):
>   Theta:=Theta1+Theta2:
>   Q:=f*fs/rho:
>   Phi:=-(f/rho)*(tau+eta*lambda^2)*fs:
>   Psi1:=(f/(2*rho))*(1-(b2-s^2)*(tau+eta*lambda^2))*fss:


>   Q:=simplify(Q);
```

$$Q := -\frac{\frac{\partial}{\partial s}\,\phi(s)}{-\phi(s) + s\left(\frac{\partial}{\partial s}\,\phi(s)\right)}$$

```
>   Theta:=simplify(Theta);
```

$$\Theta := \frac{1}{2}\,\frac{-s\,\phi(s)\left(\frac{\partial^2}{\partial s^2}\,\phi(s)\right) - s\left(\frac{\partial}{\partial s}\,\phi(s)\right)^2 + \phi(s)\left(\frac{\partial}{\partial s}\,\phi(s)\right)}{\phi(s)\left(\phi(s) - s\left(\frac{\partial}{\partial s}\,\phi(s)\right) + \left(\frac{\partial^2}{\partial s^2}\,\phi(s)\right)b2 - s^2\left(\frac{\partial^2}{\partial s^2}\,\phi(s)\right)\right)}$$

```
>   simplify(Xi/Theta-Phi/Psi1);
```

$$0$$

```
>   simplify(Xi/Theta+2*Q);
```

$$0$$

```
>   chi:=simplify(Psi1/Theta);
```

$$\chi := -\frac{(\frac{\partial^2}{\partial s^2}\,\phi(s))\,\phi(s)}{s\,\phi(s)\,(\frac{\partial^2}{\partial s^2}\,\phi(s)) + s\,(\frac{\partial}{\partial s}\,\phi(s))^2 - \phi(s)\,(\frac{\partial}{\partial s}\,\phi(s))}$$

Bibliography

[1] H. Akbar-Zadeh, *Sur les espaces de Finsler á courbures sectionnelles constantes*, Bull. Acad. Roy. Bel. Cl, Sci, 5e Série - Tome LXXXIV (1988) 281-322.

[2] P. Antonelli, R. Ingarden and M. Matsumoto, *The Theory of Sprays and Finsler Spaces with Applications in Physics and Biology*, Kluwer Academic Publishers, 1993.

[3] L. Auslander, *On curvature in Finsler geometry*, Trans. Amer. Math. Soc. **79**(1955), 378-388.

[4] D. Bao and S. S. Chern, *On a notable connection in Finsler geometry*, Houston J. Math. **19**(1)(1993), 135-180.

[5] D. Bao, S. S. Chern and Z. Shen, *An Introduction to Riemann-Finsler Geometry*, Springer, 2000.

[6] D. Bao, S. S. Chern and Z. Shen, *Rigidity issues on Finsler surfaces*, Rev. Roumaine Math. Pures Appl. **42**(1997), 707-735.

[7] D. Bao, C. Robles and Z. Shen, *Zermelo navigation on Riemannian manifolds*, J. Diff. Geom. **66**(2004), 391-449.

[8] D. Bao and C. Robles, *On Randers metrics of constant curvature*, Rep. Math. Phys. **51**(2003), 9-42.

[9] D. Bao and C. Robles, *On Ricci curvature and flag curvature in Finsler geometry*, In "*A Sampler of Finsler Geometry*" MSRI series, Cambridge University Press, 2004.

[10] D. Bao and Z. Shen, *Finsler metrics of constant curvature on the Lie group* S^3, J. London Math. Soc. **66**(2002), 453-467.

[11] S. Bácsó and M. Matsumoto, *On Finsler spaces of Douglas type. A generalization of the notion of Berwald space*, Publ. Math. Debrecen, **51**(1997), 385-406.

[12] A. Bejancu and H. R. Farran, *Finsler metrics of positive constant flag curvature on Sasakian space forms*, Hokkaido Math. J. **31**(2002), 459-468.

[13] A. Bejancu and H. R. Farran, *Randers manifolds of positive constant flag curvature*, Int. J. Math. Sci. **18**(2003), 1155-1165.

185

[14] L. Berwald, *Untersuchung der Krümmung allgemeiner metrischer Räume auf Grund des in ihnen herrschenden Parallelismus*, Math. Z. **25**(1926), 40-73.

[15] L. Berwald, *Parallelübertragung in allgemeinen Räumen*, Atti Congr. Intern. Mat. Bologna **4**(1928), 263-270.

[16] L. Berwald, *Über eine charakterische Eigenschaft der allgemeinen Räume konstanter Krümmung mit geradlinigen Extremalen*, Monatsh. Math. Phys. **36**(1929), 315-330.

[17] L. Berwald, *Über die n-dimensionalen Geometrien konstanter Krümmung, in denen die Geraden die kürzesten sind*, Math. Z. **30**(1929), 449-469.

[18] L. Berwald, *Ueber Finslersche und Cartansche Geometrie IV. Projektivkrümmung allgemeiner affiner Räume und Finslersche Räume skalarer Krümmung*, Ann. Math. **48**(1947), 755-781.

[19] R. Bryant, *Finsler structures on the 2-sphere satisfying $K = 1$*, Finsler Geometry, Contemporary Mathematics **196**, Amer. Math. Soc., Providence, RI, 1996, 27-42.

[20] R. Bryant, *Projectively flat Finsler 2-spheres of constant curvature*, Selecta Math., N.S. **3**(1997), 161-204.

[21] R. Bryant, *Some remarks on Finsler manifolds with constant flag curvature*, Houston J. Math. **28**(2) (2002), 221-262.

[22] H. Busemann, *Intrinsic area*, Ann. Math. **48**(1947), 234-267.

[23] E. Cartan, *Les espaces de Finsler*, Actualités 79, Paris, 1934.

[24] X. Chen, X. Mo and Z. Shen, *On the flag curvature of Finsler metrics of scalar curvature*, J. London Math. Soc. **68**(2) (2003), 762-780.

[25] X. Chen and Z. Shen, *Randers metrics with special curvature properties*, Osaka J. Math. **40**(2003), 87-101.

[26] X. Chen and Z. Shen, *Projectively flat Finsler metrics with almost isotropic S-curvature*, Acta Math. Sci., to appear.

[27] X. Cheng and Z. Shen, *Randers metrics of scalar flag curvature*, preprint, 2005.

[28] S. S. Chern, *On the Euclidean connections in a Finsler space*, Proc. National Acad. Soc., **29**(1943), 33-37; or Selected Papers, vol. II, 107-111, Springer, 1989.

[29] S. S. Chern, *Local equivalence and Euclidean connections in Finsler spaces*, Science Reports Nat. Tsing Hua Univ. **5**(1948), 95-121.

[30] S. S. Chern, *On Finsler geometry*, C. R. Acad. Sc. Paris **314**(1992), 757-761.

[31] S. S. Chern, *On the connection in Finsler geometry*, Chin. Ann. Math. **23B:2**(2002), 181-186.

[32] P. Dazord, *Variétés finslériennes de dimension δ-pincées*, C. R. Acad. Sc. Paris **266**(1968), 496-498.

[33] P. Dazord, *Variétés finslériennes en forme des sphéres*, C. R. Acad. Sc. Paris **267**(1968), 353-355.

[34] A. Deicke, *Über die Finsler-Räume mit $A_i = 0$*, Arch. Math. **4**(1953), 45-51.

[35] C. E. Duran, *A volume comparison theorem for Finsler manifolds*, Proc. of A.M.S. **126**(1998), 3079-3082.

[36] L. Eisenhart, *Non-Riemannian geometry*, American Mathematical Society, 1922.

[37] P. Finsler, *Über Kurven und Flächen in allgemeinen Räumen*, (Dissertation, Göttingen, 1918), Birkhäuser Verlag, Basel, 1951.

[38] P. Foulon, *Curvature and global rigidity in Finsler geometry*, Houston J. Math. **28**(2002), 263-292.

[39] P. Funk, *Über Geometrien bei denen die Geraden die Kürzesten sind*, Math. Ann. **101**(1929), 226-237.

[40] P. Funk, *Über zweidimensionale Finslersche Räume, insbesondere über solche mit geradlinigen Extremalen und positiver konstanter Krümmung*, Math. Z. **40**(1936), 86-93.

[41] G. Hamel, *Über die Geometrien in denen die Geraden die Kürzesten sind*, Math. Ann. **57**(1903), 231-264.

[42] M. Hashiguchi and Y. Ichijyō, *On some special (α, β)-metrics*, Rep. Fac. Sci. Kagoshima Univ. **8**(1975), 39-46.

[43] M. Hashiguchi and Y. Ichijyō, *Randers spaces with rectilinear geodesics*, Rep. Fac. Sci. Kagoshima Univ. (Math. Phys. & Chen.) **13**(1980), 33-40.

[44] D. Hilbert, *Mathematical Problems*, Bull. Amer. Math. Soc. **37**(2001), 407-436. Reprinted from Bull. Amer. Math. Soc. **8** (July 1902), 437-479.

[45] D. Hrimiuc and H. Shimada, *On the L-duality between Finsler and Hamilton manifolds*, Nonlinear World **3**(1996), 613-641.

[46] D. Hrimiuc and H. Shimada, *On some special problems concerning the L-duality between Finsler and Cartan spaces*, Tensor, N. S. **58**(1997), 48-61.

[47] Y. Ichijyō, *Finsler spaces modeled on a Minkowski space*, J. Math. Kyoto Univ. **16**(1976), 639-652.

[48] Y. Ichihyō, *On special Finsler connections with vanishing hv-curvature tensor*, Tensor, N. S. **32**(1978), 146-155.

[49] M. Ji and Z. Shen, *On strongly convex graphs in Minkowski geometry*, Can. Math. Bull. **45**(2) (2002), 232-246.

[50] A. B. Katok, *Ergodic properties of degenerate integrable Hamiltonian systems*, Izv. Akad. Nauk SSSR **37**(1973), [Russian]; Math. USSR-Izv. **7**(1973), 535-571.

[51] M. Kitayama, M. Azuma and M. Matsumoto, *On Finsler spaces with (α, β)-metric. Regularity, geodesics and main scalars*, J. Hokkaido Univ. Education (sect. II A), **46**(1995), 1-10.

[52] S. Kikuchi, *On the condition that a space with (α, β)-metric be locally Minkowskian*, Tensor, N.S. **33** (1979), 242–246.

[53] C.-W. Kim and J.-W. Yim, *Finsler manifolds with positive constant flag curvature*, Geom. Dedicata (to appear).

[54] W. Klingenberg, *Riemannian geometry*, de Gruyter Studies Math. **1**, 2nd rev. ed., de Gruyter Berlin New York 1995.

[55] S. Kobayashi, *Theorem of Busemann-Mayer on Finsler metrics*, Hokkaido

Math. J. **20** (1991), no. 2, 205-211.

[56] S. Kobayashi and K. Nomizu, *Foundations of differential geometry* I, II. Interscience Publishers (1963-69).

[57] D. Kosambi, *Parallelism and path-spaces*, Math. Z. **37**(1933), 608-618.

[58] D. Kosambi, *Systems of differential equations of second order*, Quart. J. Math., Oxford Ser. **6**(1935), 1-12.

[59] L. Kozma, *On Landsberg spaces and holonomy of Finsler manifolds*. Contemporary Mathematics, **196**(1996), 177-185.

[60] L. Kozma, *On holonomy groups of Landsberg manifolds*, Tensor, N. S. **62**(2000), 87-90.

[61] G. Landsberg, *Über die Totalkrümmung*, Jahresberichte der deut Math. Ver. **16**(1907), 36-46.

[62] G. Landsberg, *Über die Krümmung in der Variationsrechnung*, Math. Ann. **65**(1908), 313-349.

[63] C. LeBrun and L. J. Mason, *Zoll manifolds and complex surfaces*, J. Diff. Geom. **61**(2002), 453-535.

[64] M. Matsumoto, *On C-reducible Finsler spaces*, Tensor, N.S. **24**(1972), 29-37.

[65] M. Matsumoto, *On Finsler spaces with Randers metric and special forms of important tensors*, J. Math. Kyoto Univ. **14** (1974), 477-498.

[66] M. Matsumoto, *Foundations of Finsler Geometry and special Finsler Spaces*, Kaiseisha Press, Japan, 1986.

[67] M. Matsumoto, *Randers spaces of constant curvature*, Rep. Math. Phys. **28**(1989), 249-261.

[68] M. Matsumoto, *The Berwald connection of a Finsler space with an (α, β)-metric*, Tensor, N.S. **50**(1991), 18-21.

[69] M. Matsumoto, *Theory of Finsler spaces with (α, β)-metric*, Rep. Math. Phys. **31**(1992), 43-83.

[70] M. Matsumoto and S. Hōjō, *A conclusive theorem on C-reducible Finsler spaces*, Tensor, N. S. **32**(1978), 225-230.

[71] M. Matsumoto and H. Shimada, *The corrected fundamental theorem on Randers spaces of constant curvature*, Tensor, N. S. **63**(2002), 43-47.

[72] X. Mo, *The flag curvature tensor on a closed Finsler space*, Res. Math. **36**(1999), 149-159.

[73] X. Mo, *On the flag curvature of a Finsler space with constant S-curvature*, Houston J. Math. (to appear).

[74] X. Mo and Z. Shen, *On negatively curved Finsler manifolds of scalar curvature*, Can. Math. Bull., (to appear).

[75] X. Mo, Z. Shen and C.Yang, *Some constructions of projectively flat Finsler metrics*, preprint, 2004.

[76] S. Numata, *On Landsberg spaces of scalar curvature*, J. Korea Math. Soc. **12**(1975), 97-100.

[77] T. Okada, *On models of projectively flat Finsler spaces of constant negative curvature*, Tensor, N. S. **40**(1983), 117-123.

[78] H. B. Rademacher, *A sphere theorem for non-reversible Finsler metrics*, In *"A Sampler of Finsler Geometry"* MSRI series, Cambridge University Press, 2004.

[79] G. Randers, *On an asymmetric metric in the four-space of general relativity*, Phys. Rev. **59**(1941), 195-199.

[80] A. Rapcsák, *Über die bahntreuen Abbildungen metrischer Räume*, Publ. Math. Debrecen, **8**(1961), 285-290.

[81] C. Robles, *Einstein metrics of Randers type*, Ph.D. thesis, University of British Columbia, Canada, 2003.

[82] C. Robles, *Geodesics in Randers spaces of constant curvature*, Trans. Amer. Math. Soc. (to appear).

[83] V. S. Sabau and H. Shimada, *Classes of Finsler spaces with* (α, β)*-metrics*, Rep. Math. Phy. **47** (2001), 31-48.

[84] Z. Shen, *Finsler spaces of constant positive curvature*, In: Finsler Geometry, Contemporary Math. **196**(1996), 83-92.

[85] Z. Shen, *Volume comparison and its applications in Riemann-Finsler geometry*, Adv. Math. **128**(1997), 306-328.

[86] Z. Shen, *Differential Geometry of Spray and Finsler Spaces*, Kluwer Academic Publishers, 2001.

[87] Z. Shen, *Lectures on Finsler Geometry*, World Scientific, Singapore, 2001.

[88] Z. Shen, *Projectively flat Randers metrics of constant flag curvature*, Math. Ann. **325**(2003), 19-30.

[89] Z. Shen, *Projectively flat Finsler metrics of constant flag curvature*, Trans. Amer. Math. Soc. **355**(4) (2003), 1713-1728.

[90] Z. Shen, *Finsler metrics with K=0 and S=0*, Can. J. Math. **55**(2003), no. 1, 112-132.

[91] Z. Shen, *Two-dimensional Finsler metrics of constant flag curvature*, Manuscripta Math. **109**(3) (2002), 349-366.

[92] Z. Shen, *Nonpositively curved Finsler manifolds with constant S-curvature*, Math. Z., (to appear).

[93] Z. Shen, *Landsberg curvature, S-curvature and Riemann curvature*, In *"A Sampler of Finsler Geometry"* MSRI series, Cambridge University Press, 2004.

[94] Z. Shen and H. Xing, *On Randers metrics with isotropic S-curvature*, preprint, 2004.

[95] C. Shibata, H. Shimada, M. Azuma and H. Yosuda, *On Finsler spaces with Randers' metric*, Tensor, N. S. **31**(1977), 219-226.

[96] J. Simons, *On transitivity of holonomy systems*, Ann. Math. **76**(1962), 213-234.

[97] Z. I. Szabó, *Positive definite Berwald spaces (Structure theorems on Berwald spaces)*, Tensor, N. S. **35**(1981), 25-39.

[98] H. C. Wang, *On Finsler space with completely integrable equations of Killing*, J. London Math. Soc. **22**(1947), 5-9.

[99] J. H. C. Whitehead, *Convex regions in the geometry of paths*, Quart. J.

Math. Oxford Ser. **3**(1932), 33-42.

[100]　H. Xing, *The geometric meaning of Randers metrics with isotropic S-curvature*, Advances in Mathematics (China), (to appear).

[101]　H. Yasuda and H. Shimada, *On Randers spaces of scalar curvature*, Rep. Math. Phys. **11**(1977), 347-360.

[102]　E. Zermelo, *Über das Navigationsproblem bei ruhender oder veränderlicher Windverteilung*, Z. Argrew. Math. Mech. **11**(1931), 114-124.

[103]　W. Ziller, *Geometry of the Katok examples*, Ergod. Th & Dynam. Sys. **3**(1982), 135-157.

Index